Walch Hands-on Science Series

The ABCs of Chemistry

by Michael Margolin

illustrated by Lloyd Birmingham

Project editors: Joel Beller and Carl Raab

J. WESTON
WALCH
PUBLISHER

Portland, Maine

User's Guide
to
Walch Reproducible Books

As part of our general effort to provide educational materials that are as practical and economical as possible, we have designated this publication a "reproducible book." The designation means that purchase of the book includes purchase of the right to limited reproduction of all pages on which this symbol appears:

Here is the basic Walch policy: We grant to individual purchasers of this book the right to make sufficient copies of reproducible pages for use by all students of a single teacher. This permission is limited to a single teacher and does not apply to entire schools or school systems, so institutions purchasing the book should pass the permission on to a single teacher. Copying of the book or its parts for resale is prohibited.

Any questions regarding this policy or requests to purchase further reproduction rights should be addressed to:

Permissions Editor
J. Weston Walch, Publisher
321 Valley Street • P. O. Box 658
Portland, Maine 04104-0658

1   2   3   4   5   6   7   8   9   10

ISBN 0-8251-3931-7

Copyright © 2000
J. Weston Walch, Publisher
P. O. Box 658 • Portland, Maine 04104-0658
www.walch.com

Printed in the United States of America

Contents

 # To the Teacher

This is one in a series of hands-on science activity books for middle school and early high school students. A recent national survey of middle school students conducted by the National Science Foundation (NSF) found that:

- more than half listed science as their favorite subject.
- more than half wanted more hands-on activities.
- 90 percent stated that the best way for them to learn science was to perform experiments themselves.

The books in this series seek to capitalize on these findings. These books are not texts but supplements. They offer hands-on, fun activities that reinforce your curriculum. They will introduce learners to the methods of science by providing them with opportunities to engage in the activities of science.

The hands-on activities in this book allow students to:

- manipulate equipment.
- interpret data.
- evaluate experimental designs.
- draw inferences and conclusions.
- apply the methods of science.

This book will make chemistry more important to your learners by relating the topics to students' experiences and interests. All of these activities can be done in school, and many of them can also be done at home. Because of the nature of the chemicals and equipment required for some activities in chemistry, adult supervision is strongly recommended. Appropriate safety precautions are indicated for these activities. The author has spent his working years in the classroom, and the activities have been field tested in a public middle school and/or high school.

For this book, devoted to *The ABCs of Chemistry,* the equipment is kept simple. Every effort has been made to keep equipment costs low and to choose equipment that is readily available. When more costly equipment is suggested, alternatives are recommended so that each activity can be performed with either more or less sophisticated equipment. The activities range from the very simple (Why Are Pickles Sour?) to the difficult (How Effective Are Antacids in Controlling Stomach Acid?). There is something for every student. We strongly recommend that teachers try these activities before having students perform them.

THE ACTIVITIES CAN BE USED:

- to provide hands-on experiences pertaining to textbook content.
- to give verbally limited children a chance to succeed and gain extra credit.
- as the basis for class or school science fair projects or for other science competitions.
- to involve students in science club projects.
- as homework assignments.
- to involve parents in their children's science education and experiences.

- to inculcate an appreciation for physical science.

   Each activity has a Teacher Resource section that includes—besides helpful hints and suggestions—a scoring rubric, quiz questions, and Internet connections for those students who wish to go further and carry out the follow-up activities. Instructional objectives and the National Science Standards that apply to each activity are also provided in order for you to meet state and local expectations.

What Is the Importance of Making and Recording Observations?

 INSTRUCTIONAL OBJECTIVES

Students will be able to

- observe the properties of a burning candle.
- observe changes in the properties of a burning candle.
- distinguish between qualitative and quantitative observations.
- record observations.

NATIONAL SCIENCE STANDARDS ADDRESSED

Students demonstrate scientific inquiry and problem-solving skills by

- using physical science concepts to explain observations.
- distinguishing between qualitative and quantitative observations.
- working individually and in teams to collect and share information.
- identifying experimental variables.

Students demonstrate effective scientific communication by

- explaining scientific concepts to other students.

Students demonstrate competence with the tools and technologies of science by

- using a thermometer.
- using a metric ruler.
- recognizing sources of bias in data.
- recognizing the value of using tools to observe and measure.

MATERIALS

- Candle
- Candle holder
- Matches
- Watch or clock
- Thermometer
- Metric ruler

 = **Safety icon**

HELPFUL HINTS AND DISCUSSION

Time frame: One class period
Structure: Individuals or cooperative learning groups
Location: In class or at home

In this activity, students will learn to make and record observations. They will work on distinguishing between qualitative and quantitative observations, and they will gain appreciation for the value of using tools to make observations. It is important that students have a chance to share their observations, thus providing feedback to each other and helping to hone their observational skills. This can be accomplished by having all students or groups of students contribute to a class list of observations.

Students should be supervised by an adult when working with a burning candle.

ADAPTATIONS FOR HIGH AND LOW ACHIEVERS

High Achievers: These students should join the lower achievers in a work group and assist them. Higher-achieving students should be able to complete this activity with less supervision. An extended activity for such students might be to use a candle that doesn't go out. This would provide a situation in which students' observations are contrary to those expected. They should be encouraged to make more observations and to come to the realization that they need tools to make observations on their own. Encourage these students to perform the additional activities, particularly follow-up activities 2 and 3.

(continued)

ADAPTATIONS FOR HIGH AND LOW ACHIEVERS
(continued)

Low Achievers: Challenged students may need help and guidance in performing this activity. It is important for you to provide motivation and add greater structure to their observations. You might want to guide these students through the first few observations and, perhaps, discuss results at several points during the activity.

Introduce the idea that tools can be used to make observations. A possible introduction to tools (e.g., the ruler) is to show an optical illusion based upon size, explaining that while our senses may be fooled, a ruler is not. Another introduction to tools (e.g., the thermometer) would be to use three beakers of water—one with hot water, one with cold water, and one with warm water. Have a blindfolded student place his/her hand first in hot water, then in warm water, and describe the temperature of each. Have a second blindfolded student place his/her hand first in the cold water, then in the warm water, and describe the temperature of each. Help students recognize that their descriptions of the warm water were subjective, but a thermometer would give an objective reading—it would not be affected by the temperature of the previously sampled water.

SCORING RUBRIC
Full credit should be given to students who record observations and provide correct answers in full sentences to all the questions. Extra credit may be given for completion of any follow-up activities.

INTERNET TIE-INS   http://www.Teachnet.com/lesson/scinet.html
http://www.alincom.com/educ/sci.htm

QUIZ   1. Why do scientists often use tools to make observations?
2. Distinguish between a qualitative and a quantitative observation. Give an example of each.

Name _____   Date _____

What Is the Importance of Making and Recording Observations?

⚗ BEFORE YOU BEGIN ⚗

Two of the most important things that scientists do are making observations and recording them. Different scientists choose to observe different things, but all scientists tend to make observations of the things that interest them. For example, an astronomer may observe the behavior and properties of a star, a biologist may observe the behavior and anatomy of a frog, and a chemist may observe the properties of an acid.

If we consider the different observations that you can make, you will find that some observations *describe* things and some observations *measure* things. Observations that describe things are called **qualitative observations**. Observations that measure things are called **quantitative observations**. For example, that a frog is green and that it jumps are qualitative observations of a frog. The measured size of its legs or the exact height that it jumps are quantitative observations.

In this activity, you will observe and record the properties of a burning candle. In considering the candle and describing it, you should observe its different parts. (See the diagram at right.) You should make both qualitative and quantitative observations.

✂ MATERIALS

- Ⓢ Candle
- Candle holder
- Ⓢ Matches
- Watch or clock
- Thermometer
- Metric ruler

Ⓢ = Safety icon

PROCEDURES

Record your observations and answers to the questions in the Data Collection and Analysis section.

1. Ⓢ **This step is to be done under the direct supervision of your teacher or another adult.** Carefully light the candle. Record the time when you light it in the Data Collection and Analysis section.

2. Look at the flame. Record at least five observations about the flame in the data chart. For each observation, indicate whether it is a qualitative or quantitative observation. If you use a tool to make the observation, describe the tool in the chart.

3. Look at the wick. Record a minimum of three observations about the wick in the data chart. For each observation, indicate whether it is a qualitative or quantitative observation. If you use a tool to make the observation, describe the tool in the chart.

(continued) 🔥

What Is the Importance of Making and Recording Observations? *(continued)*

4. Look at the candle itself. Record a minimum of eight observations about the candle. For each observation, indicate whether it is a qualitative or quantitative observation. If you use a tool to make the observation, describe the tool in the chart.

5. Record the time when you blow out the candle.

DATA COLLECTION AND ANALYSIS

Time candle lit: _____

Time candle extinguished: _____

Observation	Candle Part	Qualitative / Quantitative	Tools

CONCLUDING QUESTIONS AND ACTIVITIES

1. Did any of the observations you made require you to use any senses other than vision? If so, which senses did you use? _____
_____

(continued)

What Is the Importance of Making and Recording Observations? *(continued)*

2. Review the observations that you made using tools. Which observations could have been made without a tool? List them; then list those that could not have been made without a tool. _____

_____

_____

_____

3. Why did you use tools to help you make some observations? _____

_____

_____

4. Describe a tool that you would have liked to use if it had been available. Why would this tool have been useful? _____

_____

_____

5. Review the qualitative observations that you have made. Can you make any of them quantitative? List each such qualitative observation and explain how you could make it quantitative.

_____

_____

_____

_____

Use your explanations to revise your observations, and add your new quantitative ones to the data chart.

6. Write a description of the burning candle using the observations that you made. _____

_____

_____

7. Meet with other members of your class to compare your observations. List observations you missed that other students made. _____

_____

_____

🔥 Follow-up Activities 🔥

1. Make additional observations of the burning candle and add them to the list you have developed.
2. Observe an object in your home, such as a glass of cold milk or a cup of hot coffee. Record your observations.
3. 🚫 **This activity requires care to avoid getting burned. Please check with an adult before attempting it.** Hold a metal spoon in the yellow part of a flame for approximately one minute. Record your observations of the spoon.

How Can You Tell Which Plastic Is Which?

☑ INSTRUCTIONAL OBJECTIVES

Students will be able to

- define density.
- find the density of a substance using Archimedes' principle.
- use density to identify substances.
- explain Archimedes' principle (flotation).
- observe and record data.

🌐 NATIONAL SCIENCE STANDARDS ADDRESSED

Students demonstrate an understanding of

- density.
- Archimedes' principle.
- Using a property to identify an unknown.

Students demonstrate scientific inquiry and problem-solving skills by

- using physical science concepts to explain observations.
- working individually and in teams to collect and share information.
- identifying and controlling variables in an experimental research setting.

Students demonstrate effective scientific communication by

- explaining scientific concepts to other students.
- arguing from evidence.

Students demonstrate competence with the tools and technologies of science by

- using laboratory equipment to find the density of solids using Archimedes' principle.

✂ MATERIALS

- Samples of unknown plastics, each approximately 2 cm square
- Five 500-ml beakers
- 100-ml graduated cylinder
- Tongs
- Marking pencil
- Isopropyl alcohol (rubbing alcohol)
- Stirring rod
- Distilled water
- Corn syrup
- Paper towels

HELPFUL HINTS AND DISCUSSION

Time frame: Two class periods
Structure: Individuals or cooperative learning groups of two to four students
Location: In class or at home

In this activity, students will use a physical property of a substance (density) to identify different plastics. Rather than using the usual method of finding density, they will find the density of these samples by determining whether they float or sink in liquids with different densities. Provide the students with samples of different plastics and ask them to identify each sample based on its density, which they will find by comparing its density to the known densities of common plastics. You can get these samples from hardware stores or from the school custodian. Plastic toys, utensils, and various containers (e.g., soft drink bottles, opaque juice containers, yogurt containers, medicine bottles) are good sources of plastic samples. You may also choose to substitute other substances, such as wood or laminate, for plastics. All samples should be the same size. Using a permanent marker, mark all samples of the same substance with a number. This will allow students to compare their results with class data.

ADAPTATIONS FOR HIGH AND LOW ACHIEVERS

High Achievers: These students should join the lower achievers in a work group and assist them. These students should be expected to demonstrate a fuller understanding of buoyancy. They should be encouraged to perform the additional activities, particularly follow-up activities 1 and 2.

Low Achievers: Challenged students may need a more complete explanation and possibly a demonstration of Archimedes' principle and buoyancy. They may also need help and guidance in determining the density of samples and the identity of the unknown samples from the chart of densities. Using one sample to demonstrate how to perform the activity and how to use the charts might be helpful.

SCORING RUBRIC

Full credit should be given to students who record observations and provide correct answers in full sentences to all the questions. Extra credit may be given for completion of any follow-up activities.

 **INTERNET TIE-INS**   http://charon.nmsu.edu/~maddison/A105/latex/lab/Dengrav/node1html
                                     http. Njsd.org/users/s/sherttin/phy710/genchem.labs.density.htm

 QUIZ   1. What is the density of a substance that has a mass of 5 grams and a volume of 2 ml?
        2. Explain why a substance that does not float in water might float in a solution of sugar water.

Name _____ Date _____

How Can You Tell Which Plastic Is Which?

⚗ BEFORE YOU BEGIN ⚗

You have probably observed that some things float in water while others don't. Have you ever thought about whether an object that floats in water will float in other liquids? If you think about the differences between swimming in a pool and swimming in the ocean, you probably know the answer.

An object floats in a liquid if the object's weight equals the weight of the volume of liquid that it displaces. In other words, if you get into a bathtub full of water, your body takes the place of (displaces) some of the water, which overflows onto the floor. This is not something you want to try because you will have a flooded bathroom. If the weight of the displaced water is equal to or greater than the weight of your body, you will float. An ancient Greek scientist named Archimedes was the first person to explain why things float. **Archimedes' principle** states that an object will float in a liquid if it displaces a volume of liquid equal to its weight.

Salt water is denser than fresh water, which is why it is easier to float in salt water than in a pool. If the liquid displaced were less dense than water, you would have difficulty floating.

What does it mean when someone says that a liquid is lighter or heavier than water? To compare the heaviness of different substances, you must compare equal volumes of them. Doing this is finding the **density** of the substance. The density of an object is defined as the mass per unit volume ($D = M/V$). Using densities allows you to compare the masses of equal volumes of substances.

In this activity, you will find the density of objects by seeing if they float in different liquids. If an object floats in a liquid, its density is equal to or less than the density of the liquid. If an object sinks in a liquid, its density is greater than the density of the liquid. You will test samples of several different solids in five different liquids of known density. This will allow you to determine the approximate density of these solids. You will then identify each sample by comparing its density to the densities listed in the chart of densities that follows. To perform this activity, you will need two information charts. The first chart tells you the densities of the five liquids you will make. The second lists the densities of common plastics.

DENSITIES OF SOLUTIONS

Solution/Beaker #	Density g/ml
Solution A/Beaker #1	0.91
Solution B/Beaker #2	0.93
Solution C/Beaker #3	1.00
Solution D/Beaker #4	1.16
Solution E/Beaker #5	1.36

(continued)

How Can You Tell Which Plastic Is Which? *(continued)*

DENSITIES OF COMMON PLASTICS

Symbol	Name	Density g/ml
PET	Polyethylene Terephthalate	1.38–1.39
HDPE	High Density Polyethylene	0.96–0.97
PVC	Polyvinyl Chloride	1.16–1.35
LDPE	Low Density Polyethylene	0.92–0.94
	Polypropylene	0.90–0.91
PS	Polystyrene	1.05–1.07

MATERIALS

- Samples of unknown plastics, each approximately 2 cm square
- Five 500-ml beakers
- 100-ml graduated cylinder
- Tongs
- Marking pencil
- Isopropyl alcohol (rubbing alcohol)
- Stirring rod
- Distilled water
- Corn syrup
- Paper towels

PROCEDURES

Record your observations and answers to the questions in the Data Collection and Analysis section.

1. Using the marking pencil, label the five beakers as follows: #1 (0.91); #2 (0.93); #3 (1.00); #4 (1.16); #5 (1.36).

2. Use the graduated cylinder to add 143 ml of isopropyl alcohol and 57 ml of distilled water to beaker #1. Stir the contents with the stirring rod. Rinse and dry the graduated cylinder and the stirring rod after using them.

3. Use the graduated cylinder to add 133 ml of isopropyl alcohol and 67 ml of distilled water to beaker #2. Stir the contents with the stirring rod. Rinse and dry the graduated cylinder and the stirring rod after using them.

4. Use the graduated cylinder to add 200 ml of distilled water to beaker #3. Rinse and dry the beaker after using it.

5. Use the graduated cylinder to add 100 ml of corn syrup and 100 ml of distilled water to beaker #4. Stir the contents with the stirring rod. Rinse and dry the graduated cylinder and the stirring rod after using them.

6. Use the graduated cylinder to add 150 ml of corn syrup and 50 ml of distilled water to beaker #5. Stir the contents with the stirring rod. Rinse and dry the graduated cylinder and the stirring rod after using them.

(continued)

How Can You Tell Which Plastic Is Which? *(continued)*

7. Using tongs, place the first sample into beaker #1, submerging the sample in the liquid. Observe whether it sinks or floats, and record this information in the Data Table 1. Remove the sample, rinse it in water, and dry it with a paper towel. Using the same procedure, test the same sample in beakers #2, 3, 4, and 5.

8. Repeat step 7 for each sample that your teacher gave you.

DATA COLLECTION AND ANALYSIS

Data Table 1: Enter the sample number in the first column, and indicate whether the sample floated (F) or sank (S) in each solution.

Sample Number	Sol. A (0.91 g/ml)	Sol. B (0.93 g/ml)	Sol. C (1.0 g/ml)	Sol. D (1.16 g/ml)	Sol. E (1.36 g/ml)

Data Table 2: Enter the information you discover in steps 1 and 2 below to identify each plastic.

1. Use the Densities of Solutions Chart to determine the approximate density of each plastic sample. You can do this by seeing in which solutions it sank or floated. If the sample sank, its density is greater than the solution's. If the sample floated, its density is less than the solution's. For example, if a substance sank in D and floated in E, its density is more than 1.16 (>1.16) but less than 1.36 (<1.36). In the table below, write the density range you determine for each sample using this method.

2. Use these density ranges and the information about plastics in the Densities of Common Plastics chart to determine the name of each plastic sample.

Sample Number	Density Range	Name of Plastic

(continued)

How Can You Tell Which Plastic Is Which? *(continued)*

❓ CONCLUDING QUESTIONS

1. Write a brief article explaining the technique you used to find the density of these plastics.

   _____

   _____

   _____

2. Explain why it was necessary to rinse and dry the sample between immersions in different solutions. _____

   _____

   _____

3. What steps did you take to control the variables in this experiment? _____

   _____

   _____

4. What changes in this activity would be necessary to test denser solids?_____

   _____

   _____

5. Using Archimedes' principle, explain why a solid floats in a liquid. _____

   _____

   _____

6. Compare your results to the results other groups got. _____

   _____

   _____

🧪 Follow-up Activities 🧪

1. Investigate and explain the operation of a hydrometer.
2. Use another laboratory method to determine the density of a solid.
3. Write a report on the work of Archimedes and present it to your classmates.
4. Investigate and then prepare a chart of the densities of common substances.
5. Place a can of soda and a can of diet soda of the same brand in a pail of water. Determine if they float or sink. Write a report explaining your observations. Present your findings to the class.

Gases of Burning, Part I: Oxygen

☑ INSTRUCTIONAL OBJECTIVES

Students will be able to

- generate and collect oxygen.
- observe and record data.
- collect a gas by displacing water.
- describe the properties of oxygen.

🌐 NATIONAL SCIENCE STANDARDS ADDRESSED

Students demonstrate an understanding of

- properties of oxygen.
- collection of a gas by water displacement.

Students demonstrate scientific inquiry and problem-solving skills by

- using physical science concepts to explain observations.
- using data to develop a description of oxygen.
- working in teams to collect, share, and analyze data.

Students demonstrate competence with the tools and technologies of science by

- using laboratory equipment to generate oxygen and to collect oxygen by displacing water.

✂ MATERIALS

- 🖐 Candle, approximately 6 cm long
- Jar
- Pie plate
- Water
- Gas generating bottle
- One-hole stopper
- Right-angle glass bend, each arm approximately 5 cm long
- 3 percent hydrogen peroxide (H_2O_2)
- 100-ml graduated cylinder
- Manganese dioxide (MnO_2)
- 10-ml graduated cylinder
- Pneumatic trough with tray
- Rubber tubing
- Heat-resistant pad
- Two gas collection bottles
- Two glass plates
- Wood splint
- 🖐 Matches
- Tongs
- 🖐 Steel wool (without soap)
- 🖐 Bunsen burner or alcohol burner
- Safety goggles

🖐 = Safety icon

HELPFUL HINTS AND DISCUSSION

Time frame:	Two class periods
Structure:	Cooperative learning groups of two to four students
Location:	In class

In this activity, students will generate oxygen and then investigate its properties. Oxygen will serve as an example of an element. They will generate oxygen from hydrogen peroxide (H_2O_2), observe the properties of oxygen, and observe the burning of a candle in air. Although there are several ways to prepare oxygen in a laboratory, using 3 percent H_2O_2 is the safest method for students. 🖐 **The students must wear goggles when performing this activity and should perform it under the teacher's supervision. Use scissors to cut steel wool rather than pulling it apart with your fingers to avoid skin lacerations. Caution students to be very careful and to use tongs when inserting a burning substance into oxygen.** You may wish to demonstrate the oxidation of sulfur by inserting the burning sulfur into oxygen with a deflagration spoon.

If you do not have pneumatic troughs, you can construct them from plastic containers. Trays can be made from pieces of metal. The contents of the gas generating bottle can be washed away with water and flushed down the sink.

ADAPTATIONS FOR HIGH AND LOW ACHIEVERS

High Achievers: These students should be in work groups with the lower achievers and assist them. Discuss the collection of a gas by water displacement, eliciting the properties of oxygen and water that permit the use of this collection method. If you wish to add the use of the balance to this activity and you have the equipment, you can have these students measure 4 grams of manganese dioxide (MnO_2) instead of having them use the 10-ml graduated cylinder. Encourage these students to perform the additional activities, particularly follow-up activities 2 and 3.

Low Achievers: Challenged students may need help and guidance in performing this activity. You may need to demonstrate the techniques before having students perform them. **It is particularly important to discuss and stress safety procedures.** You may choose to demonstrate the insertion of burning steel wool into oxygen, rather than have the students perform this task. Provide a glossary and/or reference material for the bold-faced terms in this activity.

SCORING RUBRIC

Full credit should be given to students who record observations and provide correct answers in full sentences to all the questions. Extra credit may be given for completion of any follow-up activities.

 **INTERNET TIE-INS**   http://scicentral.com/k-12
www.ncsu.edu/science_house/LerningMAterials/Countertop/exp7.html

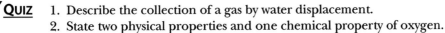 **QUIZ**   1. Describe the collection of a gas by water displacement.
2. State two physical properties and one chemical property of oxygen.

Gases of Burning, Part I: Oxygen

🧪 BEFORE YOU BEGIN 🧪

Have you ever watched someone put out a fire? Many fires are put out by cutting oxygen off from the fuel. A fuel cannot burn without oxygen. Oxygen supports burning and is one of the gases of burning. In this activity, you will investigate the properties of the *element oxygen*. To do this, you will **generate** and collect a sample of oxygen. You will generate oxygen by the **decomposition** of hydrogen peroxide (H_2O_2). You may be familiar with hydrogen peroxide, since a bottle of it can often be found in first aid kits and home medicine cabinets. Hydrogen peroxide will decompose by itself into water and oxygen. You will help this process along by using manganese dioxide (MnO_2) as a **catalyst** to speed up the decomposition of hydrogen peroxide. A catalyst is a substance that changes the speed of a chemical reaction without itself being changed.

The oxygen that you produce will be collected by a method called **water displacement**. A rubber tube will transfer the oxygen the decomposition creates into a bottle of water. The oxygen will push out (or displace) the water in the bottle. You will have collected a whole bottle of oxygen when all of the water in the bottle has been displaced. You will collect and use two bottles of oxygen to discover some of its physical and chemical properties.

✂️ MATERIALS

- 🤚 Candle, approximately 6 cm long
- Jar
- Pie plate
- Water
- Gas generating bottle
- One-hole stopper
- Right-angle glass bend, each arm approximately 5 cm long
- 3 percent hydrogen peroxide (H_2O_2)
- 100-ml graduated cylinder
- Manganese dioxide (MnO_2)
- 10-ml graduated cylinder
- Pneumatic trough with tray

- Rubber tubing
- Heat-resistant pad
- Two gas collection bottles
- Two glass plates
- Wood splint
- 🤚 Matches
- Tongs
- 🤚 Steel wool (without soap)
- 🤚 Bunsen burner or alcohol burner
- Safety goggles

🤚 = Safety icon

(continued)

Gases of Burning, Part I: Oxygen *(continued)*

PROCEDURES

Record your observations and answers to questions in the Data Collection and Analysis section.

PART I: When setting up the apparatus for this activity, refer to the diagram below.

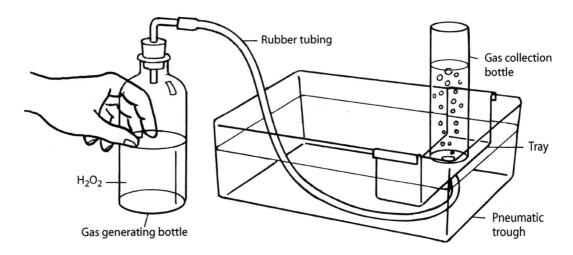

You must wear goggles while performing this activity.

1. Place the right-angle bend into the hole of the stopper. Attach the rubber tubing to the right-angle bend. Insert the stopper into the gas generating bottle.

2. Fill the pneumatic trough with enough water to cover the tray by at least 2 cm.

3. Fill one gas collection bottle with water and place a glass plate on top of it. Using both hands—one around the bottle and the other holding the glass plate in place—invert the bottle and place it on the tray of the pneumatic trough. Remove the glass plate. Repeat this step with the other gas collection bottle.

4. Attach the rubber tube so that it will deliver gas to the bottle that is directly over the hole in the tray. Do not, however, put the gas collection bottle over the hose yet.

5. Remove the stopper from the gas generating bottle and add 100 ml of hydrogen peroxide to it.

6. Use the 10-ml graduated cylinder to measure 3 ml of manganese dioxide. Add the manganese dioxide to the hydrogen peroxide in the gas generating bottle. Restopper the gas generating bottle.

7. When gas starts bubbling from the tube, let the first bubbles escape because they are air bubbles, not oxygen. After about 30 seconds, place the first gas collection bottle over the tube so that gas bubbles enter the bottle. When the bottle is full of gas, slide the glass plate under it and remove the bottle from the pneumatic trough. Stand the covered bottle upright on the table. Collect a second bottle of oxygen in the same way.

(continued)

15 *Walch Hands-on Science Series: The ABCs of Chemistry*

Gases of Burning, Part I: Oxygen (continued)

8. (✋) **Steps 8 and 9 must be done under the supervision of your teacher or another adult.** Ignite a wooden splint and then blow the flame out so that the splint is glowing. Quickly uncover one of the bottles of oxygen and carefully place the glowing splint into it. Observe what happens and record your observations in the Data Collection and Analysis section. Dip the splint in water to put it out.

9. Light the Bunsen burner. Use tongs to hold a piece of steel wool in the flame until it glows red hot. Quickly place the glowing piece of steel wool into the second bottle of oxygen. Record your observations in the Data Collection and Analysis section. Let the steel wool cool on the heat-resistant pad.

PART II: (✋) **This part must be done with your teacher or another adult supervising.**

1. Use a match to melt the bottom of the candle so it drips onto the middle of the pie plate. Fasten the candle to the pie plate with the candle drippings. Add enough water to almost fill the pie plate.

2. Light the candle and place the jar over the candle. Observe what happens and record your observations in the Data Collection and Analysis section.

DATA COLLECTION AND ANALYSIS

PART I

1. What state of matter is oxygen in at room temperature? _____

2. Does oxygen have a color? _____

3. Does oxygen have an odor? _____

4. Is oxygen soluble in water? Explain your answer. _____

   _____

5. What happened when you placed the glowing splint in the oxygen? _____

   _____

6. What happened when you placed the glowing steel wool in oxygen? _____

   _____

7. Write a brief description of the element oxygen based on your observations. _____

   _____

   _____

PART II

Describe the changes in the candle and the water level that you observed as the candle burned.

_____

_____

_____

(continued) 🔥

Gases of Burning, Part I: Oxygen *(continued)*

❓ CONCLUDING QUESTIONS

1. List the physical properties of oxygen that you observed. _____
   _____
   _____
   _____

2. Why do you think that the candle under the glass jar went out? _____
   _____
   _____

3. Why do you think that the water rose in the jar as the candle burned? Justify your answer.
   _____
   _____
   _____

4. List the chemical properties of oxygen that you observed. _____
   _____
   _____

5. Write a brief description of how to collect gas by water displacement. _____
   _____
   _____

6. Why is oxygen important to living things? _____
   _____
   _____

7. Research three methods of putting out a fire. Explain how each cuts off the supply of oxygen.
   _____
   _____
   _____

🧪 Follow-up Activities 🧪

1. All living tissue contains an enzyme that has the same effect on hydrogen peroxide as manganese dioxide. Repeat this activity using chicken liver, which is a good source of this enzyme, in place of manganese dioxide.
2. Research the production and uses of liquid oxygen.
3. Research methods of preparing other gaseous elements, such as nitrogen or hydrogen.
4. Research the properties of other gaseous elements (nitrogen, helium, hydrogen, argon, and chlorine) and construct a chart that compares their properties to those of oxygen.

Gases of Burning, Part II: Carbon Dioxide

✔ INSTRUCTIONAL OBJECTIVES

Students will be able to

- generate and collect carbon dioxide.
- observe and record data.
- collect a gas by displacing water.
- describe the properties of carbon dioxide.
- distinguish between the properties of a compound and the properties of its constituent elements.
- draw conclusions based upon data.

🌐 NATIONAL SCIENCE STANDARDS ADDRESSED

Students demonstrate an understanding of

- properties of carbon dioxide.
- collection of a gas by water displacement.
- distinction between an element and a compound.

Students demonstrate scientific inquiry and problem-solving skills by

- using physical science concepts to explain observations.
- using data to develop a description of carbon dioxide.
- working in teams to collect and share information.

Students demonstrate effective scientific communication by

- explaining scientific concepts to other students.
- representing data in multiple ways.
- arguing from evidence and data.

Students demonstrate competence with the tools and technologies of science by

- using laboratory equipment to generate carbon dioxide and to collect carbon dioxide by displacing water.

✂ MATERIALS

- 250-ml Erlenmeyer flask
- Two-hole stopper
- Thistle tube (long-stem funnel)
- Right-angle glass bend, each arm approximately 5 cm long
- 90-cm rubber tubing
- 100-ml graduated cylinder
- 10-ml graduated cylinder
- Pneumatic trough with tray
- Two gas collection bottles
- Two glass plates
- Wood splint
- 🖐 Matches
- Two 500-ml beakers
- 🖐 Candle
- 10-ml graduated cylinder
- Cardboard square, 4 cm square
- Stopper for gas collection bottle
- Test tube
- Straw
- 50 g marble chips ($CaCO_3$)
- 🖐 Lime water ($Ca(OH)_2$)
- 🖐 3M hydrochloric acid (HCl)
- Baking soda ($NaHCO_3$)
- Vinegar
- Carbon black

🖐 = Safety icon

HELPFUL HINTS AND DISCUSSION

Time frame: Two class periods
Structure: Cooperative learning groups of two to four students
Location: In class

In this activity, students will investigate the properties of a compound—carbon dioxide. They will use two methods to generate carbon dioxide and also observe the properties of carbon dioxide.

🖐 **The students must wear goggles when performing this activity and should perform this activity under the supervision of a teacher or another adult. Hydrochloric acid may cause severe burns. Lime water can irritate skin and eyes.** Therefore, working with these substances requires careful supervision and observance of appropriate safety precautions. You may wish to demonstrate the preparation of carbon dioxide from marble chips and have the students do the preparation from baking soda. *(continued)*

HELPFUL HINTS AND DISCUSSION *(continued)*

To dispose of the contents of the flask used to generate CO_2, fill it with water and flush the contents down the sink. The marble chips may be collected and dried for reuse.

If you do not have pneumatic troughs, you can construct them from plastic containers and make the trays from pieces of metal.

ADAPTATIONS FOR HIGH AND LOW ACHIEVERS

High Achievers: These students should work in groups with lower achievers and assist them. You may discuss the effect of temperature on solubility of a gas in water. Encourage these students to perform the additional activities, particularly follow-up activities 2 and 5.

Low Achievers: Challenged students may need help and guidance in performing this activity. You may need to demonstrate the techniques before having students perform them. **It is particularly important to discuss safety procedures.** Provide a glossary and/or reference material for the bold-faced terms in this activity.

SCORING RUBRIC

Full credit should be given to students who record reasonable observations and provide correct answers in full sentences to all the questions. Extra credit may be given for completion of any follow-up activities.

 **INTERNET TIE-INS**   www.edf.org/Want2Help/b_gw20steps.html
http://wwitch.unl.edu/safety/hslab7.html
www.ncsu.edu/science_house/LearningMaterialFolder/CountertopChem/exp9.html

QUIZ   1. How does carbon dioxide differ from its component elements?
2. Describe two physical properties and one chemical property of carbon dioxide.

Gases of Burning, Part II: Carbon Dioxide

⚗️ BEFORE YOU BEGIN ⚗️

Carbon dioxide is considered a gas of burning because it results from burning. While oxygen is always a **reactant** of burning, carbon dioxide is a **product** of burning. In this activity, you will investigate the properties of the compound carbon dioxide and compare them to the properties of the elements that make it up. The formula for carbon dioxide is **CO$_2$**. The formula tells you that a molecule of carbon dioxide is made of two atoms of the element oxygen and one atom of the element carbon.

This compound is important for many reasons. Carbon dioxide is one of the products of **respiration** and one of the reactants in **photosynthesis**. This makes CO$_2$ very important in the study of biology. In addition, carbon dioxide is produced by burning such fuels as coal or oil. Carbon dioxide is often referred to as a **greenhouse gas**. This means that adding it to the atmosphere locally helps to increase the temperature of the atmosphere globally. For these and other reasons, carbon dioxide is a very important compound.

✂️ MATERIALS

- 250-ml Erlenmeyer flask
- Two-hole stopper
- Thistle tube (long-stem funnel)
- Right-angle glass bend, each arm approximately 5 cm long
- 90-cm rubber tubing
- 100-ml graduated cylinder
- 10-ml graduated cylinder
- Pneumatic trough with tray
- Two gas collection bottles
- Two glass plates
- Wood splint
- (✋) Matches
- Two 500-ml beakers

- (✋) Candle
- 10-ml graduated cylinder
- Cardboard square, 4 cm square
- Stopper for gas collection bottle
- Test tube
- Straw
- 50 g marble chips (CaCO$_3$)
- (✋) Lime water (Ca(OH)$_2$)
- (✋) 3M hydrochloric acid (HCl)
- Baking soda (NaHCO$_3$)
- Vinegar
- Carbon black

(✋) = **Safety icon**

(continued)

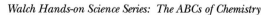

Gases of Burning, Part II: Carbon Dioxide *(continued)*

PROCEDURES

Record your observations and answers to questions in the Data Collection and Analysis section.

PART I

1. Measure 5 ml of baking soda in the 10-ml graduated cylinder. Pour the baking soda into a beaker.

2. Use the graduated cylinder to measure 50 ml of vinegar. Add the vinegar to the beaker. Cover the beaker with a glass plate. Observe what happens and record your observations in the Data Collection and Analysis section. You should see a gas bubbling. The gas you have produced is carbon dioxide.

3. Under the direct supervision of your teacher or another adult, use a match to melt the bottom of the candle so that it drips onto the middle of the cardboard square. Fasten the candle there with the candle drippings. Place the cardboard and candle inside a beaker. Light the candle.

4. Bring your beaker of carbon dioxide near the beaker with the candle. Remove the glass plate and pour the carbon dioxide into the beaker holding the candle. Don't pour any of the liquid into the beaker. Observe what happens and record your observations in the Data Collection and Analysis section. Try to relight the candle. Were you successful?

PART II: When setting up the apparatus for this activity, refer to the diagram below.

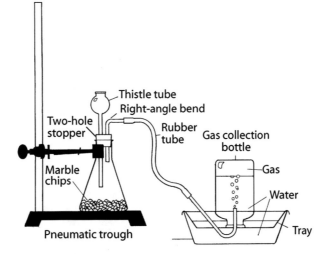

You must wear goggles while performing this activity.

1. **This step must be performed under the supervision of your teacher or another adult.** Assemble the apparatus as pictured in the diagram. Place the right-angle bend into one hole of the stopper and place the thistle tube into the other. Attach the rubber hose to the right-angle bend.

2. Place the marble chips into the flask. Place the stopper securely into the mouth of the flask, making sure that the bottom of the thistle tube is about 5 cm above the bottom of the flask.

(continued)

Gases of Burning, Part II: Carbon Dioxide *(continued)*

3. Through the thistle tube, add to the flask 100 ml of water from the graduated cylinder.

4. Fill the pneumatic trough with water.

5. Fill the two gas collection bottles with water. Place a glass plate on top of each bottle. Invert them and place them, one at a time, on top of the tray in the pneumatic trough. Remove the glass plates.

6. Attach the rubber tubing to the pneumatic trough so that it will deliver gas to one of the bottles, but do not put a gas collection bottle directly over the hose yet.

7. 🖐 **This step must be performed under the supervision of your teacher or another adult.** Through the thistle tube add 6 ml of acid to the beaker. Handle the acid carefully. Fizzing will occur where the acid comes in contact with the marble chips.

8. When gas has started bubbling from the tubing, place the first gas collection bottle over it so that gas bubbles into the bottle. When the bottle is full, slide the glass plate under it. Remove the bottle and glass plate from the pneumatic trough. Stand the covered bottle on the table right side up. Collect a second bottle of carbon dioxide in the same way. Add more acid if the bubbling slows down.

9. 🖐 **This step must be performed under the supervision of your teacher or another adult.** Ignite a wooden splint and place it into the first bottle of carbon dioxide. Record your observations in the Data Collection and Analysis section.

10. Pour 25 ml of lime water into the second bottle of carbon dioxide. Stopper the bottle and **swirl** it. Observe any changes that occur. Continue to swirl the bottle. Record your observations in the Data Collection and Analysis section.

11. Add 25 ml of lime water to a test tube. Use a straw to blow into the lime water and describe your observations in the Data Collection and Analysis section.

12. Look at a sample of carbon and record your observation about the properties of carbon in the Data Collection and Analysis section.

13. Review your data from Activity 3 and add your observations about the properties of oxygen to the Data Collection and Analysis section of this activity.

DATA COLLECTION AND ANALYSIS

PART I

1. Describe what happened when you added vinegar (acetic acid) to the baking soda (sodium bicarbonate). _____

_____

2. Describe what happened when you poured carbon dioxide over the burning candle. _____

_____

_____

3. Describe what happened when you tried to relight the candle. _____

_____

(continued)

Gases of Burning, Part II: Carbon Dioxide *(continued)*

PART II

1. What state of matter is carbon dioxide in at room temperature? _____
   _____

2. Does carbon dioxide have a color? _____

3. Does carbon dioxide have an odor? _____

4. Is carbon dioxide soluble in water? _____

5. What happened when you placed the burning splint in the carbon dioxide? _____
   _____

6. What happened when you added lime water to the carbon dioxide? _____
   _____

7. What happened when you blew through the straw into lime water? _____
   _____

8. What are the properties of carbon? _____
   _____

9. What are the properties of oxygen? _____
   _____

10. Based on your observations, write a brief description of the properties of carbon dioxide.
    _____
    _____
    _____

❓ CONCLUDING QUESTIONS

1. Why do you think that the candle went out when you poured carbon dioxide over it? _____
   _____
   _____

2. Why were you able to pour the carbon dioxide? Justify your answer. _____
   _____
   _____

3. List the physical properties of carbon dioxide that you observed. _____
   _____
   _____

4. List the chemical properties of carbon dioxide that you observed. _____
   _____
   _____

(continued)

Gases of Burning, Part II: Carbon Dioxide *(continued)*

5. Which do you think is denser, carbon dioxide or air? Justify your answer. _____

_____

_____

6. Which gas (oxygen or carbon dioxide) is present in your breath? Justify your answer. _____

_____

_____

_____

7. Complete the chart below comparing carbon dioxide, carbon, and oxygen.

Properties of Carbon	Properties of Oxygen	Properties of Carbon Dioxide

8. Write a statement that compares the properties of carbon dioxide to the elements that make it up. _____

_____

_____

9. Why do you think that carbon dioxide is used in some fire extinguishers? _____

_____

_____

⚗ Follow-up Activities ⚗

1. Research and write a report about the production and use of dry ice.
2. Research and write a report that compares the properties of table salt (sodium chloride) to the properties of sodium and chlorine.
3. Research and write a report about the use of baking soda in baking.
4. Research and write a report about the effect of carbon dioxide on global warming.
5. Research and construct a model of a CO_2 fire extinguisher.

Making Soap

✔ INSTRUCTIONAL OBJECTIVES

Students will be able to

- describe saponification.
- record observations.

🌐 NATIONAL SCIENCE STANDARDS ADDRESSED

Students demonstrate an understanding of

- saponification.

Students demonstrate scientific inquiry and problem-solving skills by

- using physical science concepts to explain observations.
- working individually or in teams to collect and share information.

Students demonstrate effective scientific communication by

- explaining scientific concepts to other students.

Students demonstrate competence with the tools and technologies of science by

- using a hot-water bath.
- setting up and performing a filtration.

✂ MATERIALS

- Three 500-ml beakers
- Two test tubes
- Test-tube holder
- Test-tube rack
- Tripod
- 🖐 Bunsen burner/alcohol heater
- 🖐 Matches
- 25-ml graduated cylinder
- 50-ml graduated cylinder
- Ring stand
- Ring
- Funnel
- Filter paper
- Scoopula
- Stirring rod
- Wash bottle
- 🖐 5M sodium hydroxide (NaOH)
- Vegetable shortening
- Sodium chloride
- Ethanol
- Water
- Goggles
- Protective gloves
- Laboratory apron

🖐 = **Safety icon**

HELPFUL HINTS AND DISCUSSION

Time frame: Two class periods
Structure: Individuals or cooperative learning groups
Location: In class

In this activity, students will make soap. Saponification is the reaction between a base and a fat or oil to produce soap and glycerol. The nature of the fat or oil used will determine the properties of the soap. If the students use a fat that is solid at room temperature (indicating a higher molecular weight), the soap will be solid.

🖐 **Because they are working with a strong base, students must wear goggles, and the teacher or another adult must supervise this activity. Because sodium hydroxide is very caustic, it must be handled safely.**

ADAPTATIONS FOR HIGH AND LOW ACHIEVERS

High Achievers: These students should join lower achievers in a work group and assist them. Discuss the use of the hot water bath. Encourage these students to perform the additional activities, particularly follow-up activities 2 and 3.

Low Achievers: Provide a glossary and/or reference material for the bold-faced terms in this activity. Review safety precautions with these students.

SCORING RUBRIC

Full credit should be given to students who record observations and provide correct answers in full sentences to all the questions. Extra credit may be given for completion of any follow-up activities.

INTERNET TIE-INS   http://candleandsoap.miningco.com
www.ballarat.net.au/~standeyo/News/INFO_Files/soapmaking.html

QUIZ   1. Explain why many soap samples have a pH above 7.
2. Why must you wear goggles when making soap?

Name _____   Date _____

Making Soap

⚗ BEFORE YOU BEGIN ⚗

You use soap in various forms every day—to wash your hands, your clothes, and your dishes. Think of all the activities that use soap. Did you ever wonder where soap comes from? Is it a naturally occurring substance, or is it something people make? The process of making soap is called **saponification**. In this activity, you will make soap.

Saponification is a chemical reaction between a **lipid** (fat or oil) and a base. The soap that forms is actually a type of compound called the **salt** of an organic acid. Many soaps are called lye soaps because the base that is used is sodium hydroxide, which is also known as lye. The word equation for saponification is:

lipid + sodium hydroxide = soap + glycerol

🖐 **The base that you will use in this reaction, sodium hydroxide, is very caustic and can burn your skin and eyes. You must be very careful in handling this substance. Always wear goggles, protective gloves, and a laboratory apron while performing this activity, and be sure to wash any spills with water.**

✂ MATERIALS

- Three 500-ml beakers
- Two test tubes
- Test-tube holder
- Test-tube rack
- Tripod
- 🖐 Bunsen burner/alcohol heater
- 🖐 Matches
- 25-ml graduated cylinder
- 50-ml graduated cylinder
- Ring stand
- Ring
- Funnel

- Filter paper
- Scoopula
- Stirring rod
- Wash bottle
- 🖐 5M sodium hydroxide (NaOH)
- Vegetable shortening
- Sodium chloride
- Ethanol
- Water
- Goggles
- Protective gloves
- Laboratory apron

🖐 = Safety icon

📐 PROCEDURES

Record your observations and answers to questions in the Data Collection and Analysis section.

1. Fill one of the beakers approximately half full of water and place it on top of the tripod.

2. 🖐 **For the rest of this activity, you must wear goggles and work under the supervision of your teacher or another adult.**

   Place the Bunsen burner or alcohol heater under the tripod and safely light it. Adjust the burner to produce a blue flame. When the water boils, the beaker of boiling water will be a **hot-water bath**.

(continued) 🔥

© 2000 J. Weston Walch, Publisher   27   Walch Hands-on Science Series: The ABCs of Chemistry

Making Soap *(continued)*

3. Use the scoopula to transfer enough vegetable shortening to fill the test tube about one third full.

4. Using the test-tube holder, transfer the test tube to the hot-water bath. Leave the test tube in the bath until all of the shortening melts.

5. Using the test-tube holder, remove the test tube from the beaker and place it in the test-tube rack.

6. Use the 25-ml graduated cylinder to transfer 20 ml of sodium hydroxide **very carefully** to the test tube.

7. Now transfer 5 ml of ethanol to the test tube, using the graduated cylinder.

8. Using the test-tube holder, return the test tube to the hot-water bath. Continue heating the water so that it boils slowly. Leave the test tube in the hot-water bath for 25 minutes. Stir the contents of the test tube every two to three minutes. See the diagram at above right.

9. While the soap is forming, measure 45 ml of sodium chloride in the 50-ml graduated cylinder. Transfer the sodium chloride to another beaker containing approximately 200 ml of water. Stir thoroughly until the sodium chloride is dissolved. Then, set up the filtering apparatus shown in the diagram at below right.

10. Carefully pour the contents of the test tube into the beaker of salt water. The soap should form **curds** (lumps) of soap in the beaker.

11. Filter the contents of the beaker. When all of the liquid has passed through the filter, wash the soap with water, using the wash bottle. Let the soap dry on the filter paper.

12. Using the scoopula, scrape up some of your soap and transfer it to a clean test tube.

13. Add 25 ml of water to the test tube and shake vigorously. Describe your observations in the Data Collection and Analysis section.

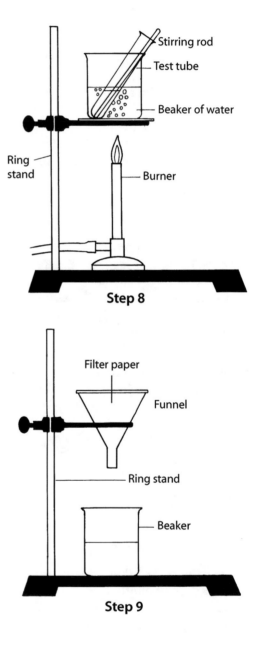

Step 8

Step 9

(continued)

Name _____ Date _____

Making Soap *(continued)*

DATA COLLECTION AND ANALYSIS

1. Describe what happened when you shook the test tube containing your soap and water.

   _____

   _____

2. Describe your safety precautions during this activity and the reason for each.

   _____

   _____

   _____

CONCLUDING QUESTIONS

1. Why did you use a hot-water bath in this activity rather than an open flame? _____

   _____

   _____

2. Why did you wash the soap after filtering it? _____

   _____

3. Write a brief article explaining how soap is made. _____

   _____

   _____

   _____

🔺 Follow-up Activities 🔺

1. Research and write a report about commercial soap making. Present the report to the class.
2. Research the effect of hard water on soap. Suggest an experiment to test the effect of hard water on soap.

Making a Simple Battery

☑ INSTRUCTIONAL OBJECTIVES

Students will be able to

- identify that a chemical cell (battery) produces electricity by a chemical reaction.
- observe and record data.
- construct a battery.

🌐 NATIONAL SCIENCE STANDARDS ADDRESSED

Students demonstrate an understanding of

- chemical cells and batteries.

Students demonstrate scientific inquiry and problem-solving skills by

- using physical science concepts to explain observations.
- working in teams to collect and share information.

Students demonstrate effective scientific communication by

- explaining scientific concepts to other students.

Students demonstrate competence with the tools and technologies of science by

- using laboratory equipment to construct a battery.

✂ MATERIALS

- Baby-food jar
- One-hole rubber stopper (#9)
- Strip of copper metal
- Magnesium strip
- Steel wool
- 15-cm length of dialysis tubing
- Two wires with alligator clips
- Flashlight bulb in socket
- Metric ruler
- 0.5 M sodium sulfate (Na_2SO_4)
- 0.5 M copper(II) sulfate ($CuSO_4$)
- Funnel

HELPFUL HINTS AND DISCUSSION

Time frame: One class period
Structure: Cooperative learning groups of two to four students
Location: In class

In this activity, students will construct a chemical cell (battery) and will use it to light a flashlight bulb. You may also wish to connect six of these cells to operate a 9-volt toy.

ADAPTATIONS FOR HIGH AND LOW ACHIEVERS

High Achievers: These students should be in working groups with lower achievers and assist them. Encourage them to perform the additional activities, particularly follow-up activities 1 and 2.

Low Achievers: Challenged students may need help in assembling the cell and should be supervised by you or another adult.

SCORING RUBRIC

Full credit should be given to students who correctly construct the cell and provide correct answers in full sentences to all the questions. Extra credit may be given for completion of any follow-up activities.

💻 INTERNET TIE-INS

www.exploratorium.com/snacks/hand_battery.html
http://library.advanced.org/10784/chem2.html
www.washingtonpost.com/wp-srv/interact/longterm/horizon/010897/battery.htm

❓ QUIZ

1. Describe how a chemical reaction can produce electricity.
2. What is a battery?

Making a Simple Battery

⚗ BEFORE YOU BEGIN ⚗

If you have ever turned on a flashlight, awakened to a traveling alarm clock, listened to a boom box, or used any electric device that runs without being plugged in, you have used a battery. A **battery**, used as a source of electric energy, is actually a combination of **chemical cells**. A chemical cell can also be referred to as a **voltaic cell**. It uses a chemical reaction to produce an **electric current**. In a chemical cell, one substance loses electrons to another substance that gains them. The loss of electrons will produce positive ions. The gain of electrons will produce negative ions. This flow of electrons between substances is called an electric current. The first person to observe that a chemical reaction could produce an electric current was Alessandro Volta in the 1790s. The *volt* and *voltaic cell* are named in his honor.

Some batteries, such as the one you will make, require liquids and are called "wet cells." Other batteries, named "dry cells," use dry substances. The flashlight battery is an example of a dry cell. Many batteries eventually run down and are thrown away. Some batteries, however, can be recharged. In a rechargeable battery, a source of electricity is used to reverse the chemical reaction. These batteries store energy that is released as electricity when needed.

In this activity, you will construct a chemical cell that uses magnesium and copper. When the magnesium reacts with sodium sulfate, it will give up electrons. The copper ions in the copper sulfate solution will then gain these electrons to form solid copper.

$$Mg \longrightarrow Mg^{2+} + 2e \qquad Cu^{2+2e} \longrightarrow Cu$$

After you and your classmates have constructed these chemical cells, your teacher can show you how to put them together to make a battery.

 **MATERIALS**

- Baby-food jar
- One-hole rubber stopper (#9)
- Strip of copper metal
- Magnesium strip
- Steel wool
- 15-cm length of dialysis tubing

- Two wires with alligator clips
- Flashlight bulb in socket
- Metric ruler
- 0.5 M sodium sulfate (Na_2SO_4)
- 0.5 M copper(II) sulfate ($CuSO_4$)
- Funnel

(continued)

Making a Simple Battery (continued)

PROCEDURES

Record your observations and answers to questions in the Data Collection and Analysis section. Refer to the following diagrams as you assemble the battery.

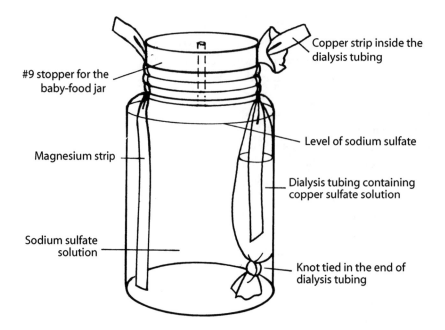

#9 stopper for the baby-food jar

Copper strip inside the dialysis tubing

Level of sodium sulfate

Magnesium strip

Dialysis tubing containing copper sulfate solution

Sodium sulfate solution

Knot tied in the end of dialysis tubing

1. Use the steel wool to clean the strips of copper and magnesium until they are shiny.

2. Hold the piece of dialysis tubing under warm running water until it becomes flexible and the tubing opens. Then, tie a knot in one end to form a bag.

3. Using the funnel, fill the bag with copper (II) sulfate solution, leaving the top 5 to 6 cm empty. Insert the copper strip into the bag, as shown in the diagram above. Place the bag into the baby food jar, bending the top of the bag and the copper strip over the top edge of the jar.

4. Place the magnesium strip in the jar (on the opposite side from the bag containing the copper strip), and bend the top of the strip over the top edge of the jar to hold the strip in place.

5. Fill the jar with sodium sulfate solution, leaving the neck of the jar empty.

6. Insert the stopper so that the metal strips and bag are held in place.

7. Use the alligator clips to attach the wire leads to the two metal strips.

8. Attach the wire leads to the socket containing the light bulb. Describe what happens when you connect the battery to the light bulb in the Data Collection and Analysis section.

(continued)

Name _____   Date _____

Making a Simple Battery *(continued)*

STUDENT ACTIVITY PAGE

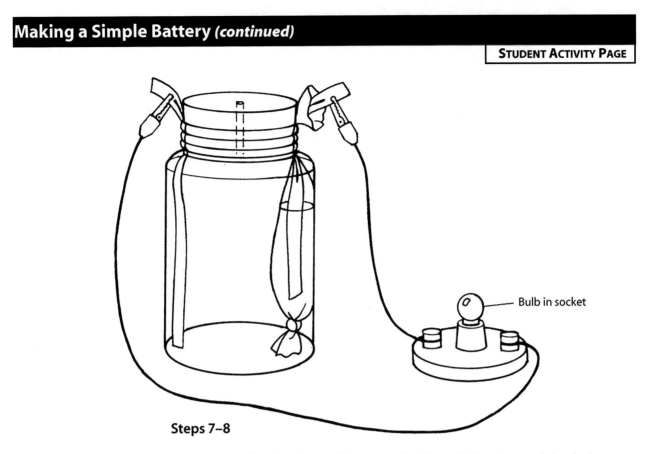

Steps 7–8

9. Observe any changes to the metal strips. Record them in the Data Collection and Analysis section.

DATA COLLECTION AND ANALYSIS

1. What happens when you attach the battery to the light bulb? Describe it fully. _____

    _____

    _____

    _____

    _____

2. Describe any changes you observe in the metal strips. _____

    _____

    _____

    _____

    _____

(continued)

© 2000 J. Weston Walch, Publisher   33   *Walch Hands-on Science Series: The ABCs of Chemistry*

Making a Simple Battery *(continued)*

❓ CONCLUDING QUESTIONS

1. What reasons can you give for the green coating that forms on the magnesium strip? _____

   _____

   _____

   _____

2. Do you expect the bulb to continue glowing forever? Explain your answer. _____

   _____

   _____

   _____

3. In which directions do electrons move in this reaction? _____

   _____

   _____

4. Describe how chemical cells can be combined to make a battery. _____

   _____

   _____

   _____

5. Write a paragraph explaining how a chemical cell works. _____

   _____

   _____

   _____

   _____

   _____

   _____

⚗ Follow-up Activities ⚗

1. Research and write a report about car storage batteries. Present this report to your teacher.
2. Research and write a report explaining how dry cells work. Present this report to your class.
3. Investigate how watch batteries work. Compare them to other types of batteries.
4. Research and write an article for your class about the work of Alessandro Volta.

Identifying Unknowns

✔️ INSTRUCTIONAL OBJECTIVES

Students will be able to

- use solubility to identify cations in known and unknown solutions.
- observe and record data.
- define the terms cation and precipitate.

🌐 NATIONAL SCIENCE STANDARDS ADDRESSED

Students demonstrate an understanding of

- qualitative analysis.
- precipitates.
- certain properties of matter.

Students demonstrate scientific inquiry and problem-solving skills by

- using physical science concepts to explain observations.
- working individually and in teams to collect and share information.
- evaluating the design of an investigation.

Students demonstrate effective scientific communication by

- explaining scientific concepts to other students.
- arguing from data to identify unknowns.

Students demonstrate competence with the tools and technologies of science by

- using laboratory equipment.
- using techniques of qualitative analysis to identify unknowns.

✂️ MATERIALS

- 10 test tubes
- Test-tube rack
- Test-tube holder
- 10-ml graduated cylinder
- 100-ml graduated cylinder
- 300-ml beaker
- Medicine droppers
- Marking pencil
- Stirring rod
- Ring stand and ring
- Wire gauze
- 🖐 Bunsen burner
- 🖐 Matches
- Distilled water

- Goggles
- 0.1 M silver nitrate ($AgNO_3$)
- 0.1 M mercury (I) nitrate ($Hg_2(NO_3)_2$)
- 0.1 M lead (II) nitrate ($Pb(NO_3)_2$)
- 0.1 M sodium nitrate ($NaNO_3$)
- 0.1 M calcium nitrate ($Ca(NO_3)_2$)
- 🖐 1.0 M hydrochloric acid (HCI)
- 1 M potassium chromate (K_2CrO_4)
- 5 M ammonia (NH_3)
- 0.1 M ammonium carbonate (NH_4CO_3)
- Unknown samples 1–5

🖐 = Safety icon

HELPFUL HINTS AND DISCUSSION

Time frame: Two class periods
Structure: Individuals or cooperative learning groups of two to four students
Location: In class

In this activity, students will determine the precipitates that form when different anions are added to five cations: NA^+, Ca^{2+}, Ag^+, Hg_2^{2+}, and Pb^{2+}. They will learn to distinguish between these cations by observing their reactions with specific anions. They will use this information to determine whether or not these cations are present in unknown solutions. The first four unknowns will contain only one cation, and your students will determine which of two possible cations is present. You may choose to give some students an unknown containing one of these cations and some students an unknown containing the other. This will prevent students from copying results. If you do this, code the unknowns with a two-digit code, for example, 1a and 1b. The fifth unknown will contain two cations, and students will be asked to plan a method for determining which cations are present. The unknown samples should be distributed in labeled vials, each containing 5 ml of the unknown. **It is important to stress the necessity of cleaning glassware before using it, since contamination with any solution would alter the results.** Explain that *cations* are pronounced *cat-ions*. You may wish to have students perform this activity over a two-day period.

ADAPTATIONS FOR HIGH AND LOW ACHIEVERS

High Achievers: These students should work with lower achievers and assist them. You may choose to make the identification of cations in the fifth unknown more difficult by telling students that the unknown may contain one through four cations. Explain to these students that a negative test result is as important as a positive test result. Encourage these students to perform the additional activities, particularly follow-up activities 1 and 4.

Low Achievers: Challenged students may need help performing these activities and planning their scheme for identifying unknowns. For these students, you may wish to limit the list of possible cations in the fifth unknown to three rather than five.

SCORING RUBRIC

Give full credit to students who record correct observations, correctly identify unknowns, and provide correct answers in full sentences to all the questions. Extra credit may be given for completion of any follow-up activities.

INTERNET TIE-INS   http://comptons3.aol.com/encyclopedia/ARTICLES/00949_A.html
www.wqa.org/Glossery/index.html
www.scimedia.com/chem-ed/analytic/
www.shef.ac.uk/chemistry/web-elements/main/index-nofr.html
www.lapeer.lib.mi.us/Chem/Chem1Docs/IdentSoln.html

QUIZ   1. Describe the steps you would take to determine whether an unknown solution contains silver ions or lead ions.
2. Define a precipitate.

Identifying Unknowns

⚗ BEFORE YOU BEGIN ⚗

Homeowners may want to know if minerals are present in their drinking water. A scientist may want to know what minerals are in lake water or what gases are in the air we breathe. A detective may want to know whether a person has taken drugs or been given poison. All of these people would turn to an analytic chemist for answers to these questions. Analytic chemistry is a branch of chemistry that deals with identifying the presence and quantities of chemicals.

In this activity, you will become an analytic chemist. Solutions may contain positively charged ions that are called **cations**. Cations may combine with negative ions (**anions**) to form insoluble compounds. When these insoluble compounds form, they drop out of solution as solids, known as **precipitates**. A precipitate is a solid compound that forms when two ions combine in a solution. Different cations form precipitates with different anions. In this activity, you will learn some of the precipitates that form and then use this information to identify the presence of different cations in a solution.

✂ MATERIALS

- 10 test tubes
- Test-tube rack
- Test-tube holder
- 10-ml graduated cylinder
- 100-ml graduated cylinder
- 300-ml beaker
- Medicine droppers
- Marking pencil
- Stirring rod
- Ring stand and ring
- Wire gauze
- ✋ Bunsen burner
- ✋ Matches

- Distilled water
- Goggles
- 0.1 M silver nitrate ($AgNO_3$)
- 0.1 M mercury (I) nitrate ($Hg_2(NO_3)_2$)
- 0.1 M lead (II) nitrate ($Pb(NO_3)_2$)
- 0.1 M sodium nitrate ($NaNO_3$)
- 0.1 M calcium nitrate ($Ca(NO_3)_2$)
- ✋ 1.0 M hydrochloric acid (HCl)
- 1 M potassium chromate (K_2CrO_4)
- 5 M ammonia (NH_3)
- 0.1 M ammonium carbonate (NH_4CO_3)
- Unknown samples 1–5

✋ = Safety icon

📦 PROCEDURES

✋ **You must wear goggles while performing this activity.**

PART I

1. Using the marking pencil, label five test tubes as follows: Na^+, Ca^{2+}, Ag^+, Hg_2^{2+}, Pb^{2+}. Place these test tubes in the test-tube rack. Label the remaining five test tubes U1 through U5 and place them aside for use in Part II.

(continued)

Identifying Unknowns *(continued)*

2. Use the graduated cylinder to add 5 ml of each of the following to the appropriate test tube:
   - sodium nitrate to the test tube labeled Na^+
   - calcium nitrate to the test tube labeled Ca^{2+}
   - silver nitrate to the test tube labeled Ag^+
   - mercury (I) nitrate to the test tube labeled Hg_2^{2+}
   - lead (II) nitrate to the test tube labeled Pb^{2+}.

   Use a clean medicine dropper to add 10 drops of hydrochloric acid to each of the five test tubes. Indicate the test tubes where precipitates formed and the color of each precipitate in the Data Collection and Analysis section. **It is important that you carefully and thoroughly wash the graduated cylinder after each use.**

3. Separate the test tubes containing precipitates from those that did not form precipitates.

4. Set up a hot water bath as follows:

5. Light the Bunsen burner and heat the water in the beaker to create a hot-water bath.

6. **Decant** the liquid in each of the three test tubes by pouring off the liquid that remains after each precipitate forms. Use the stirring rod to make sure you don't lose any of the chloride precipitate.

7. Wash the graduated cylinder carefully and use it to add 20 ml of distilled water to each of the three test tubes containing precipitates. Place these three test tubes into the hot-water bath. Leave them there for five minutes.

8. While the test tubes are in the hot-water bath, work with the two test tubes where no precipitates formed. Use a clean medicine dropper to add 5 drops of ammonium carbonate to each of the two test tubes. In the Data Collection and Analysis section, indicate the test tube where a precipitate formed and its color. Indicate also where no precipitate formed.

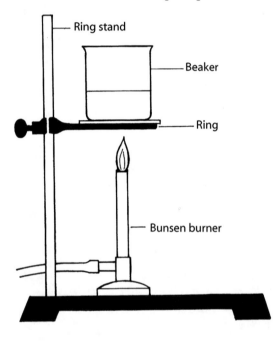

Ring stand

Beaker

Ring

Bunsen burner

9. Turn off the Bunsen burner. Use the test-tube holder to remove the three test tubes from the hot-water bath and place them in the test-tube rack. Let the test tubes cool.

10. After the test tubes have cooled, place rubber stoppers into each one and shake it. Observe which precipitates dissolve in hot water. Indicate this in the Data Collection and Analysis section.

11. Use a clean medicine dropper to add 5 drops of 1 M potassium chromate to the test tube labeled Pb^{2+}. Describe the precipitate that forms.

12. Use a clean medicine dropper to add 10 drops of ammonia to each of the remaining two test tubes. Describe what happens in each test tube in the Data Collection and Analysis section.

(continued)

Identifying Unknowns (continued)

PART II

For each unknown tested, identify the cation(s) present and explain how you determined this identity in the Data Collection and Analysis section.

1. Obtain the unknown samples from your teacher.
2. Pour Unknown 1 into the test tube labeled U1. This sample contains either Ca^{2+} or Na^+. Identify which cation is present in the sample and explain your reasoning.
3. Pour Unknown 2 into the test tube labeled U2. This sample contains either Ag^+ or Na^+. Identify which cation is present in the sample and explain your reasoning.
4. Pour Unknown 3 into the test tube labeled U3. This sample contains either Ag^+ or Pb^{2+}. Identify which cation is present in the sample and explain your reasoning.
5. Pour Unknown 4 into the test tube labeled U4. This sample contains either Ag^+ or Hg_2^{2+}. Identify which cation is present in the sample and explain your reasoning.
6. Pour Unknown 5 into the test tube labeled U5. This sample contains two cations. Formulate and explain a plan for identifying the cations present. Identify the cations present.

DATA COLLECTION AND ANALYSIS

PART I

1. When hydrochloric acid is added to the test solutions, any precipitates that form are called chlorides. Indicate below where precipitates formed and their color.

Cation Tested	Results
Na^+	
Ca^{2+}	
Ag^+	
Hg_2^{2+}	
Pb^{2+}	

2. Indicate below the results of heating chloride precipitates.

Cation Present	Result of Heating the Precipitate
Ag^+	
Hg_2^{2+}	
Pb^{2+}	

(continued)

Identifying Unknowns (continued)

3. Indicate below the result of adding ammonium carbonate.

Cation Tested	Result of Adding $(NH_4)_2CO_3$
Ca^{2+}	
Na^+	

4. What is the result of adding potassium chromate to Pb^{2+} ions? _____

_____

5. Indicate the result of adding ammonia to Ag^+ and Hg_2^{2+}.

Cation Tested	Result of Adding Ammonia
Ag^+	
Hg_2^{2+}	

PART II

1. The cation present in Unknown 1 is _____. I was able to identify it because:

_____

_____

2. The cation present in Unknown 2 is _____. I was able to identify it because:

_____

_____

3. The cation present in Unknown 3 is _____. I was able to identify it because:

_____

_____

4. The cation present in Unknown 4 is _____. I was able to identify it because:

_____

_____

5. The plan I would use to identify the cations in Unknown 5 is as follows: _____

_____

_____

6. The cations in Unknown 5 are: _____

_____

_____

(continued)

Identifying Unknowns (continued)

CONCLUDING QUESTIONS

1. Ag^+, Pb^{2+}, and Hg_2^{2+} ions form precipitates when chloride ions (Cl^-) are added to them. What compounds form? _____

   _____

   _____

2. When you added potassium chromate to the Pb^{2+} ions, which ion combined with the Pb^{2+} ions?

   _____

   _____

3. Why was it important to wash the glassware before each use? _____

   _____

   _____

4. Write a brief article explaining how you can identify the presence or absence of cations in a solution. _____

   _____

   _____

   _____

5. How would you verify that a solution does not contain lead (II) ions? _____

   _____

   _____

⚗ Follow-up Activities ⚗

1. Devise a plan for identifying the cations in an unknown solution that contains four cations. Write a description of your plan and submit it to your teacher for approval before implementing it.
2. Research and write a report on the use of a spectroscope in identifying unknowns. Present your report to the class.
3. Research and write a report on the use of flame tests to identify cations. Give the report to your teacher.
4. Plan an experiment to test water samples for the presence of cations.

How Wet Is Wet?

✔ INSTRUCTIONAL OBJECTIVES

Students will be able to

- record observations.
- define the term *surfactant*.
- describe the effect of surfactants.
- draw conclusions from observations.

🌐 NATIONAL SCIENCE STANDARDS ADDRESSED

Students demonstrate an understanding of

- properties of matter.

Students demonstrate scientific inquiry and problem-solving skills by

- using physical science concepts to explain observations.
- evaluating the outcome of the investigation.
- working individually or in teams to collect and share information.
- identifying and controlling experimental variables.

Students demonstrate effective scientific communication by

- arguing from evidence.

✂ MATERIALS

- Five 500-ml beakers
- Samples of liquid detergent, shampoo, liquid soap, liquid fabric softener
- Five similar strips of cotton cloth, approximately 6 cm by 2 cm
- Water
- Stirring rod
- 10-ml graduated cylinder
- 500-ml graduated cylinder
- Stopwatch or watch with second hand

HELPFUL HINTS AND DISCUSSION

Time frame: One class period
Structure: Individuals or cooperative learning groups
Location: In class or at home

This exercise permits students to explore the effects of surfactants on cotton fabric and to compare the effects of different surfactants on the same fabric. The color or pattern of the fabric does not matter, as long as all samples used by one student or group of students is taken from the same sample. Although students should learn to use appropriate laboratory glassware, you may need to modify the required equipment for students to do this exercise at home, where they are unlikely to have beakers and graduated cylinders. You might suggest measuring the volume of water with a measuring cup and of samples with a measuring spoon. **Emphasize that the design of the experiment requires accurately measuring these volumes.** If students are unfamiliar with volumetric measurement, demonstrate the use of the graduated cylinder.

In this activity, students will be asked to draw conclusions from their observations. Be careful to explain the difference between a conclusion based on observations and an inference or generalization that interprets the observations, as students tend to generalize from limited experimental evidence. In this activity, they can conclude that one of the liquids is more effective as a surfactant than another, but they cannot generalize about all liquids in that category or about the quantities of surfactants in the liquids.

ADAPTATIONS FOR HIGH AND LOW ACHIEVERS

High Achievers: These students should join lower achievers in a work group and assist them. Discuss the experimental design of this activity, stressing the importance of controlling experimental variables. Review the meaning of control, as well as independent and dependent variables. Encourage these students to perform the additional activities, particularly follow-up activities 1 and 2.

Low Achievers: Provide a glossary and/or reference material for the bold-faced terms in this activity. Discuss experimental design with these students, explaining the importance of controlling experimental variables.

SCORING RUBRIC

Give full credit to students who record observations and provide correct answers in full sentences to all the questions. Extra credit may be given for completion of any follow-up activities.

 **INTERNET TIE-INS**   www.surfactant.com
www.buscom.com/archive
www.teachnet.com/lesson/scimet.html
www.nmsi.ac.uk/on_line/challenge/index.html

QUIZ   1. Explain why detergents and other water-soluble cleansers contain surfactants.
2. Why is it important for a chemist to measure the volume of liquids used in an experiment?

Name _____ Date _____

How Wet Is Wet?

🧪 BEFORE YOU BEGIN 🧪

Very often, physical or biological factors limit what we can do. For example, I would like to keep a box of homemade cookies next to my bed for the next month. Unfortunately, if I do, the cookies will get stale and may become moldy. Chemists develop ways of getting around some of these limitations. For example, chemists create food additives that extend the time that baked goods will stay fresh.

In this activity, you will explore how a group of chemicals called **surfactants** affect water. Many fabrics are water resistant, which is very nice if you are trying to stay dry in the rain. But, water-resistant fabric is a problem if you want to wash it. Surfactants are chemicals developed to help solve this problem. Water normally has a high **surface tension**, the quality of a liquid's surface that helps it to form droplets rather than simply spreading out. Surfactants lower the surface tension of water, thereby allowing water to wet fabrics that would ordinarily repel it. You will observe the behavior of various surfactants and compare their effectiveness in different liquids.

You will be asked to draw conclusions from your observations. Remember that your conclusions can be based only on the data collected in this activity. You cannot extend your conclusions beyond what you have observed. You may suggest a hypothesis that generalizes your observations. A hypothesis should be a clear statement that can be confirmed by experimentation.

✂️ MATERIALS

- Five 500-ml beakers
- Samples of liquid detergent, shampoo, liquid soap, liquid fabric softener
- Five similar strips of cotton cloth, approximately 6 cm by 2 cm
- Water
- Stirring rod
- 10-ml graduated cylinder
- 500-ml graduated cylinder
- Stopwatch or watch with second hand

🔷 PROCEDURES

Record all of your observations and answers to questions in the Data Collection and Analysis section.

1. Place the five beakers on your desk and label them: *water, detergent, shampoo, liquid soap, fabric softener.*

2. Use the 500-ml graduated cylinder to add 250 ml of water to each of the beakers.

3. Using the 10-ml graduated cylinder, add 10 ml of water to the beaker labeled *water.* Stir the contents of the beaker with the stirring rod.

4. Thoroughly wash the 10-ml graduated cylinder and stirring rod. Then, use them to add 10 ml of detergent to the beaker labeled *detergent.* Stir the contents of the beaker.

5. Again, thoroughly wash the 10-ml graduated cylinder and stirring rod. Then, use them to add 10 ml of shampoo to the beaker labeled *shampoo.* Stir the contents of the beaker.

(continued)

How Wet Is Wet? *(continued)*

6. After washing the 10-ml graduated cylinder and stirring rod once again, use them to add 10 ml of liquid soap to the beaker labeled *liquid soap*. Stir the contents of the beaker.

7. Thoroughly wash the graduated cylinder and stirring rod once again. Then, use them to add 10 ml of fabric softener to the beaker labeled *fabric softener*.

8. Place a strip of cloth in the beaker labeled *water*. Start the timer and observe the piece of cloth for three minutes. Record any changes you observe during that time, such as the sinking of the cloth.

9. Place a strip of cloth in the beaker labeled *detergent*. Start the timer and observe the piece of cloth until it sinks. Record the time it took for the cloth to sink to the bottom of the beaker.

10. Place a strip of cloth in the beaker labeled *shampoo*. Start the timer and observe the piece of cloth until it sinks. Record the time it took for the cloth to sink to the bottom of the beaker.

11. Place a strip of cloth in the beaker labeled *liquid soap*. Start the timer and observe the piece of cloth until it sinks. Record the time it took for the cloth to sink to the bottom of the beaker.

12. Place a strip of cloth in the beaker labeled *fabric softener*. Start the timer and observe the piece of cloth until it sinks. Record the time it took for the cloth to sink to the bottom of the beaker.

DATA COLLECTION AND ANALYSIS

Record the time it took for the fabric to sink in each of the mixtures. If the strip of cloth did not sink, write "did not sink" in the time column.

Contents of Beaker	Time to Sink (in Seconds)
1. Water	
2. Detergent	
3. Shampoo	
4. Liquid soap	
5. Fabric softener	

1. In which beaker did the cloth sink fastest? _____

   _____

2. In which beaker did the cloth sink most slowly? _____

   _____

3. In which beaker did the cloth not sink? _____

   _____

(continued)

How Wet Is Wet? *(continued)*

❓ CONCLUDING QUESTIONS

1. Write a brief statement describing any conclusions you reached from this experiment. Support your statement with data from this experiment. _____

   _____

   _____

   _____

   _____

2. Using your data, can you determine which mixture contained the most surfactant or the most effective surfactant? Explain your answer. _____

   _____

   _____

   _____

   _____

3. Why was it necessary to wash the graduated cylinder and stirring rod before reusing them?

   _____

   _____

   _____

4. Why was it necessary to add 10 ml of water to the first beaker? _____

   _____

   _____

5. What steps did you take to limit the variables in this experiment? _____

   _____

   _____

⚗️ Follow-up Activities ⚗️

1. Design an experiment comparing the effectiveness of the surfactants in different brands of detergents.
2. Design an experiment that shows the effect of a detergent on different types of fabrics.
3. Place 5 ml of olive oil into a test tube of water, shake it, and observe the oil after several minutes. Repeat this procedure with 5 ml of detergent added to the test tube. Compare the appearance of the oil in the two test tubes.

How Much Foam Can You Make?

✔ INSTRUCTIONAL OBJECTIVES

Students will be able to

- identify the limiting reactant in a chemical reaction.
- explain that the limiting reactant determines the amount of product produced.
- identify the properties of a foam.
- observe and record data.
- construct a bar graph.

🌐 NATIONAL SCIENCE STANDARDS ADDRESSED

Students demonstrate an understanding of

- limiting reactant in a reaction.
- foams.

Students demonstrate scientific inquiry and problem-solving skills by

- using physical science concepts to explain observations.
- working individually and in teams to collect and share information.
- identifying experimental variables.

Students demonstrate effective scientific communication by

- explaining scientific concepts to other students.
- representing data in multiple ways.

Students demonstrate competence with the tools and technologies of science by

- using laboratory equipment to perform a controlled experiment.
- recognizing sources of bias in data.

✂ MATERIALS

- Reaction cylinder made of clear plastic, 1–1.5 l
- 100-ml graduated cylinder
- 10-ml graduated cylinder
- Stirring rod
- Cylinder brush
- Metric ruler or tape measure
- Sodium bicarbonate ($NaHCO_3$)
- Laundry detergent
- Distilled water
- Vinegar (10 percent acetic acid)
- Four different-colored crayons

HELPFUL HINTS AND DISCUSSION

Time frame: Two class periods
Structure: Individuals or cooperative learning groups of two to four students
Location: In class or at home

In this activity, students will investigate the formation of a foam and the effect of changing the concentration of each reactant on the amount of product generated. They will discover which is the limiting reactant and the effect of increasing its concentration. Students will determine the amount of foam produced by measuring the height of the column of foam. This activity will also broaden students' understanding of a controlled experiment.

The cylinder for this activity should be a clear plastic cylinder that will hold between 1000 ml and 1500 ml of liquid. It does not have to be a graduated cylinder, since the students will be measuring the height of the foam column, not its volume. The stirring rod must be long enough to work in the cylinder.

If this activity is done by groups of students, they can divide up responsibilities and take turns performing them. It is important that students thoroughly clean the cylinder between uses so that no reactant residue remains from one trial to the next.

In this activity, the reaction between acetic acid and sodium bicarbonate generates carbon dioxide. It is the dispersal of CO_2 in the detergent solution that produces the foam, which is a colloid.

ADAPTATIONS FOR HIGH AND LOW ACHIEVERS

High Achievers: These students should work with lower achievers and assist them. They should be able to complete this activity with less supervision and should be expected to demonstrate a better understanding of the controlling variables and identifying limiting reactants. Encourage these students to perform the additional activities, particularly follow-up activities 1 and 2.

Low Achievers: Challenged students may need help and guidance in performing these activities. You may need to review with them how to construct and read a bar graph. They may also need help to recognize that the concentration of acetic acid is the limiting factor in this reaction. Stress the concept that the less time the reaction takes to complete, the faster it occurs.

SCORING RUBRIC

Full credit should be given to students who record observations and provide correct answers in full sentences to all the questions. Extra credit may be given for completion of any follow-up activities.

INTERNET TIE-INS   http://pc65.frontier.osrhe.edu/hs/science/cmass2.htm
www.lsc.ucdavis.edu./JimSTEP/LIMITING_Rs.htm
www.styrofoam.net.au/float.htm

QUIZ   1. Why does using an electric mixer to whip egg whites produce more foam than whipping the egg whites by hand?
2. What is meant by a limiting reactant in a chemical reaction?

How Much Foam Can You Make?

⚗ BEFORE YOU BEGIN ⚗

If you have ever taken a bubble bath, eaten a lemon meringue pie, drunk from a Styrofoam cup, or used shaving cream, you have used a **foam**. A foam is a mixture made of a gas dispersed in a liquid or solid. Foams belong to a larger category of mixtures called **colloids**. A colloid is made up of small particles of one substance (the **dispersed phase**) suspended in another substance (the **continuous phase**). Because the particles in a colloid are very small, they tend to stay suspended in the continuous phase for a long time. Mixtures with large particles don't stay suspended, and the particles settle to the bottom. Generally, if the label on a mixture says to shake before using, it is a suspension, not a colloid.

In this activity, you will make a foam by releasing carbon dioxide gas into a solution of laundry detergent. The CO_2 gas will be generated by combining sodium bicarbonate with vinegar, as in Activity 4. You will also determine which of the reactants (sodium bicarbonate or acetic acid) controls the amount of carbon dioxide produced. Finally, you will perform a controlled experiment to determine which of the reactants limits the amount of foam that can be produced.

✂ MATERIALS

- Reaction cylinder made of clear plastic, 1–1.5 l
- 100-ml graduated cylinder
- 10-ml graduated cylinder
- Stirring rod
- Cylinder brush
- Metric ruler or tape measure

- Sodium bicarbonate ($NaHCO_3$)
- Laundry detergent
- Distilled water
- Vinegar (10 percent acetic acid)
- Four different-colored crayons

◈ PROCEDURES

Record your observations and answers to questions in the Data Collection and Analysis section.

1. Use the 10-ml graduated cylinder to measure 5 ml of sodium bicarbonate. Add it to the large reaction cylinder. Wash and dry the graduated cylinder. In similar fashion, add 5 ml of laundry detergent to the large reaction cylinder. Record the amounts in the data table.

2. Then, use the 100-ml graduated cylinder to add 25 ml of distilled water to the reaction cylinder. Record this amount in the data table. Stir the contents of the reaction cylinder thoroughly with the stirring rod.

3. Use the 100-ml graduated cylinder again, this time to add 25 ml of distilled vinegar to the reaction cylinder. Record this amount in the data table.

4. This mixture should begin to foam. When the foam has finished forming, measure the height of the foam column starting at the base of the cylinder. Record this measurement in the data table.

5. Using the brush, thoroughly wash the reaction cylinder so that no residue remains. **This step is very important!**

(continued) 🔥

How Much Foam Can You Make? *(continued)*

6. Repeat steps 1 through 5 using 10 ml of sodium bicarbonate, 5 ml of laundry detergent, 25 ml of water, and 25 ml of vinegar. Remember to enter the results of each trial in the data table.

7. Repeat steps 1 through 5 using 5 ml of sodium bicarbonate, 10 ml of laundry detergent, 25 ml of water, and 25 ml of vinegar.

8. Repeat steps 1 through 5 using 5 ml of sodium bicarbonate, 5 ml of laundry detergent, 50 ml of water, and 25 ml of vinegar.

9. Repeat steps 1 through 5 using 5 ml of sodium bicarbonate, 5 ml of laundry detergent, 25 ml of water, and 50 ml of vinegar.

10. In each of the above steps, you doubled the amount of one of the reactants: sodium bicarbonate, laundry detergent, water, or vinegar. **The greatest amount of foam was produced by doubling the amount of _____.**

11. Repeat steps 1 through 5, but triple the amount of the reactant identified in step 10.

DATA COLLECTION AND ANALYSIS

DATA TABLE

Trial Number	Amount of $NaHCO_3$	Amount of Laundry Detergent	Volume of Water	Volume of Vinegar	Height of Foam Column (cm)
1.	5 ml	5 ml	25 ml	25 ml	
2.					
3.					
4.					
5.					
6.					

1. Use the figure on the next page to construct a bar graph that shows how different amounts of reactant affect the height of the foam column. Use a different-colored crayon to color each column up to the height that matches the amount of foam produced. In column 1 you will indicate the height of the column in trial 1, in column 2, the height for trial 2, and so forth.

(continued)

How Much Foam Can You Make? *(continued)*

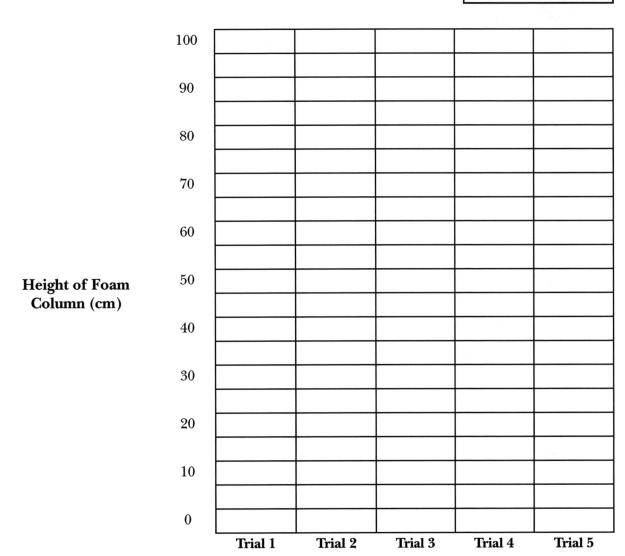

Height of Foam Column (cm)

Trial 1   Trial 2   Trial 3   Trial 4   Trial 5

CONCLUDING QUESTIONS

1. Which of the reactants had the greatest effect on the amount of foam produced?

   _____

2. Which of the reactants had the least effect on the amount of the foam produced?

   _____

3. Write a word equation for the reaction that produced carbon dioxide.

   _____

(continued)

How Much Foam Can You Make? *(continued)*

4. Was the amount of foam produced determined by the amount of detergent available or by the amount of carbon dioxide? Justify your answer. _____

_____

_____

_____

_____

5. Which reactant was the limiting factor in the production of foam? Justify your answer. _____

_____

_____

_____

_____

6. Why did you change only one reactant in each test trial? _____

_____

_____

7. What did you learn from this experiment? _____

_____

_____

8. Write a brief paragraph that describes the properties of a foam.

_____

_____

_____

_____

_____

🧪 Follow-up Activities 🧪

1. Devise an experiment to determine the maximum amount of vinegar that will affect the amount of foam produced.
2. Devise an experiment to determine if different brands of laundry detergent produce different amounts of foam.
3. Investigate and write a report on the nature of colloids. Present your findings to your classmates.
4. Research and write a report for your teacher on the commercial use of foams.

INSTRUCTIONAL OBJECTIVES

Students will be able to

- identify the properties of a colloid.
- identify the properties of a non-Newtonian fluid.
- observe and record observations.

NATIONAL SCIENCE STANDARDS ADDRESSED

Students demonstrate an understanding of

- non-Newtonian fluids.
- colloids.

Students demonstrate scientific inquiry and problem-solving skills by

- using physical science concepts to explain observations.
- working individually or in teams to collect and share information.

Students demonstrate effective scientific communication by

- explaining scientific concepts to other students.

Students demonstrate competence with the tools and technologies of science by

- using laboratory equipment to produce a colloid.

MATERIALS

- Newspaper
- 100-ml graduated cylinder
- 10-ml graduated cylinder
- 25-ml graduated cylinder
- Stirring rod or spoon
- Mixing bowl
- Teaspoon
- 500-ml beaker
- Food coloring
- Cornstarch
- Water
- Liquid laundry starch

- Sodium chloride (table salt)
- White glue
- Paper towels

HELPFUL HINTS AND DISCUSSION

Time frame: Two class periods
Structure: Individuals or cooperative learning groups
Location: At home or in class

This activity can be easily done at home since all of the materials and equipment are often found there. In this activity, students will make two different nontoxic colloids and observe their properties. You may choose to have students make only one of the colloids or divide the class into two groups, each group making one colloid. These colloids can be stored in reclosable plastic bags. Spills can be washed up with water, and the colloids thrown out with the trash.

ADAPTATIONS FOR HIGH AND LOW ACHIEVERS

High Achievers: These students should work with lower achievers and assist them. Encourage these students to perform the additional activities, particularly follow-up activities 1 and 2.

Low Achievers: Provide a glossary and/or reference material for the bold-faced terms in this activity. These students may need supervision, since this activity can be messy if not carefully done.

SCORING RUBRIC

Full credit should be given to students who record observations and provide correct answers in full sentences to all the questions. Extra credit may be given for completion of any follow-up activities.

 INTERNET TIE-INS
www.ncsu.edu/science_house/LearningMaterials Folder/CountertopChem/
www.exploratorium.edu/science_explorer/ooze.html
www.Slais.ubc.ca/users/lassa/odue/slime3.htm
http://personal.cfw.com/~rollinso/SciPhys.html

QUIZ
1. List the properties of a non-Newtonian fluid.
2. What are the properties of a colloid?

Name _____   Date _____

Why Won't the Ketchup Pour?

⚗️ BEFORE YOU BEGIN ⚗️

Have you ever had difficulty pouring ketchup from a bottle? You probably tried hitting the bottom of the bottle and found nothing came out. Ketchup is an example of a **non-Newtonian fluid**. Sir Isaac Newton was a famous eighteenth century physicist, best known for his work on gravity. He also described the properties of fluids. Water is a good example of a liquid that behaves like a Newtonian fluid. Ketchup and quicksand are two examples of non-Newtonian fluids that don't behave like water.

In this activity, you will produce two **colloids**, which are non-Newtonian fluids. A colloid is a mixture that is made up of particles suspended in a fluid. The particles are too large to form a solution, but small enough to remain suspended in the solvent. You will explore the properties of these fluids, learn how to get ketchup out of a bottle, and find out what to do if you ever fall into quicksand.

✂️ MATERIALS

- Newspaper
- 100-ml graduated cylinder
- 10-ml graduated cylinder
- 25-ml graduated cylinder
- Stirring rod or spoon
- Mixing bowl
- Teaspoon
- 500-ml beaker

- Food coloring
- Cornstarch
- Water
- Liquid laundry starch
- Sodium chloride (table salt)
- White glue
- Paper towels

PROCEDURES

Record your observations and answers to questions in the Data Collection and Analysis section.

PART I

1. Spread out the newspaper on the tabletop and work on it to keep the tabletop clean.
2. Use the 100-ml graduated cylinder to put 250 ml of cornstarch into the mixing bowl.
3. Add two drops of food coloring to the cornstarch.
4. Use the graduated cylinder to measure 125 ml of water and slowly add the water to the cornstarch, a little at a time. Mix the cornstarch and water together with a stirring rod until all the cornstarch is wet.
5. Continue to add water slowly and mix until the cornstarch has the consistency of thick ketchup. This is "slime"!

(continued)

Why Won't the Ketchup Pour? *(continued)*

6. Scoop up the mixture with your hands and roll it into a ball. Perform the operations in Part I of the Data Collection and Analysis section to the ball of "slime" that you have made and describe its properties.

PART II

7. Place 30 ml of white glue into a beaker.

8. In another beaker, add 3 ml of sodium chloride, then 60 ml of liquid laundry starch, and stir thoroughly.

9. Pour the sodium chloride-starch mixture into the beaker of glue and stir.

10. Over the sink or a large bowl, pour the mixture into your hand and squeeze out the extra liquid. Be careful not to drop this colloid.

11. Knead the colloid until if forms a smooth ball.

12. Perform the operations in Part I of the Data Collection and Analysis section to your colloid ball and describe its properties.

DATA COLLECTION AND ANALYSIS

PART I

(a) Holding the ball in your hand, gently tap it with a spoon.

(b) Holding the ball in your hand, tap it forcefully.

(c) Let the ball stay in your hand for several minutes; then, quickly squeeze the ball.

(d) Put the ball back into the mixing bowl and stir it with the stirring rod or spoon.

(e) Bounce the ball on the table.

(f) Stretch the ball.

(g) Pull the stretched ball quickly.

(h) Reform the ball and place it on the table. Leave the ball on the table for several minutes and then observe its shape.

(i) Press the ball onto a piece of newspaper, lift it, and observe the part that was pressed onto the newspaper.

PART II

(a) Based on the above tests, describe the properties of the "slime" that you made in Part I.

_____

(b) Based on the above tests, describe the properties of the colloid that you made in Part II.

_____

(c) Compare the effect of stirring upon each of the substances you made.

_____

_____

(continued)

Why Won't the Ketchup Pour? *(continued)*

❓ CONCLUDING QUESTIONS

1. Prepare a chart comparing the properties of your two colloids.
2. How are these fluids like a liquid? _____
   _____
3. How are these fluids like a solid? _____
   _____
4. How did these colloids differ? _____
   _____
5. Viscosity is a fluid's resistance to flowing freely. Which of the two colloids was most viscous? Explain your answer. _____
   _____
6. Which of the two colloids was most like ketchup? _____
7. Based on what you learned in this activity, why do people have difficulty pouring ketchup?
   _____
   _____
8. Quicksand is a non-Newtonian fluid that gets more viscous when force is applied to it. Based on this activity, what should a person who falls into quicksand do to get out? Explain your answer.
   _____
   _____
   _____
9. You would like to sell one of your colloids as a children's toy. Write a commercial to sell this product. _____
   _____
   _____

⚗ Follow-up Activities ⚗

1. Research and write a report for the class about the properties of colloids.
2. Pour about 500 ml of ketchup into a 600-ml beaker. Drop a steel ball into the ketchup from a height of 3–4 cm and time how long it takes to reach the bottom. Stir the ketchup for 1–2 minutes and repeat the ball drop. Compare the time it takes before and after stirring the ketchup for the ball to reach the bottom. Explain the results of your investigation.
3. Research and write a report for the class about thixotropy.
4. Research and write a report about dilatancy.

How Can We Separate the Parts of a Mixture?

 **INSTRUCTIONAL OBJECTIVES**

Students will be able to

- identify the properties of a mixture.
- demonstrate the techniques of decantation, filtration, and evaporation.
- observe and record data.
- distinguish between homogeneous and heterogeneous mixtures.
- explain that the physical properties of mixture materials can be used to separate them.

NATIONAL SCIENCE STANDARDS ADDRESSED

Students demonstrate an understanding of

- separation of the components of a mixture based on properties of the components.
- properties of homogeneous and heterogeneous mixtures.

Students demonstrate scientific inquiry and problem-solving skills by

- using physical science concepts to explain observations.
- working individually and in teams to collect and share information.
- identifying experimental variables.

Students demonstrate effective scientific communication by

- representing data in multiple ways.

Students demonstrate competence with the tools and technologies of science by

- using laboratory equipment to decant a liquid, filter a mixture, and recrystallize by evaporating a solvent.
- recognizing sources of bias in data.

MATERIALS

- Sodium chloride (table salt)
- Potting soil
- Two 250-ml beakers
- Two 100-ml beakers
- 25-ml graduated cylinder
- Stirring rod
- Funnel
- Filter paper
- Ring stand and ring
- Evaporating dish
- Tripod and wire gauze
- Gas burner or alcohol heater
- Matches
- Forceps

= Safety icon

HELPFUL HINTS AND DISCUSSION

Time frame: Two class periods
Structure: Cooperative learning groups of two to four students
Location: In class

In this activity, students will investigate the formation of a heterogeneous mixture and the use of various techniques to separate its components. If your school has limited resources and lab equipment is not available, you may modify some of the required equipment. For example, you can make a stand to hold the funnel or to support the evaporating dish from a wire coat hanger. See the following diagram. You can substitute plastic drinking cups for the beakers, use a coffee filter, and substitute a small Pyrex™ dish for the evaporating dish. **You or another adult should supervise the evaporating process and any other process involving heat. Students should wear safety goggles when performing this part of the activity.**

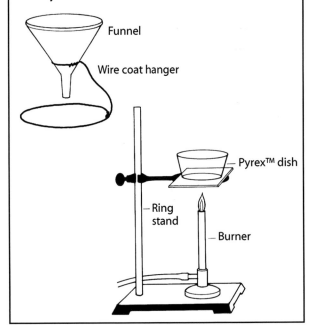

57

ADAPTATIONS FOR HIGH AND LOW ACHIEVERS

High Achievers: These students should work with lower achievers and assist them. Encourage these students to perform the additional activities, particularly follow-up activities 2 and 3.

Low Achievers: Challenged students may need help and guidance in performing these activities. Demonstrate the techniques associated with decantation, filtration, and evaporation.

SCORING RUBRIC

Full credit should be given to students who record observations and provide correct answers in full sentences to all the questions. Extra credit may be given for completion of any follow-up activities.

INTERNET TIE-INS   http://www.a/incom.com/edu/sci.htm
http://sciencemag.org

QUIZ   1. How can you distinguish between a homogenous mixture and a heterogeneous mixture?
2. How can filtration be used to separate particles suspended in a water sample?

How Can We Separate the Parts of a Mixture?

⚗ **BEFORE YOU BEGIN** ⚗

In this activity, you will separate the different materials in a mixture by using the properties of each material. You will observe samples of potting soil and salt that have been given to you. Next, you will combine them into a mixture. Then, you will use the properties you observed to separate them. There are two kinds of mixtures. In **homogeneous** mixtures all the components are uniformly mixed. For example, a glass of soda or a cup of coffee is a homogeneous mixture. In **heterogeneous** mixtures the parts can be easily separated. Separating the parts of a mixture is a task chemists often face. For example, purifying drinking water involves many of the techniques you will explore in this activity.

✂ MATERIALS

- Sodium chloride (table salt)
- Potting soil
- Two 250-ml beakers
- Two 100-ml beakers
- 25-ml graduated cylinder
- Stirring rod
- Funnel
- Filter paper

- Ring stand and ring
- Evaporating dish
- Tripod and wire gauze
- ✋ Gas burner or alcohol heater
- ✋ Matches
- Forceps

 = Safety icon

PROCEDURES

Record all of your observations and answers to questions in the Data Collection and Analysis section.

1. You have been given samples of table salt (NaCl) and potting soil. Using the 25-ml graduated cylinder, place 15 ml of soil into one small beaker and 15 ml of salt into the other small beaker. Be sure to clean and dry the graduated cylinder after measuring the soil. Observe each of these and list their properties in Data Table 1. Add a small amount of water to each, and add your observations to the list in Data Table 1.

2. Combine 15 ml of soil and 5 ml of salt in the 250-ml beaker. Describe the mixture you have made. Is it a homogeneous mixture or a heterogeneous mixture?

3. Add enough water to half fill the beaker. Use the stirring rod to mix the mixture. In Data Table 2, describe the appearance of the mixture with the water added.

4. Let the mixture stand until most of the soil has settled. Using the technique pictured in Figure 1, **decant** the liquid so that most of the soil remains behind. To decant means to pour off a liquid while leaving suspended solids behind.

(continued)

How Can We Separate the Parts of a Mixture? *(continued)*

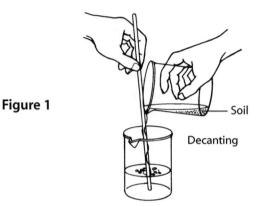

Figure 1

Soil

Decanting

5. Dispose of the remaining soil in the trash can. **Do not pour it down the sink.** Wash and dry the beaker, which you will use later.

6. Set up the ring stand on your desk.

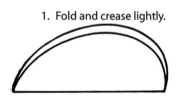

Figure 2

Ring stand

7. Fold the piece of filter paper in half and then into quarters. Open up the folded filter paper to form a cone, as shown in Figure 3. Place it into the funnel. Place the funnel into the ring of the ring stand.

Figure 3

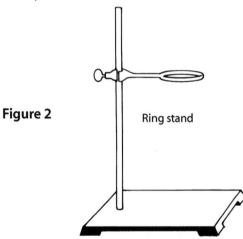

1. Fold and crease lightly.

2. Fold again.

3. Tear off corner.

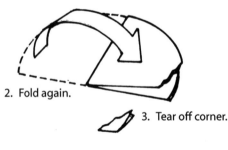
4. Open out like this.

How to fold filter paper

(continued)

How Can We Separate the Parts of a Mixture? *(continued)*

8. Place a 250-ml beaker below the funnel. Lower the ring so that the funnel is just above the beaker. Pour the decanted liquid into the funnel, as shown in Figure 4.

Figure 4

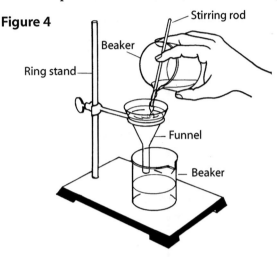

- Stirring rod
- Beaker
- Ring stand
- Funnel
- Beaker

9. The liquid that you collect in the beaker is called the **filtrate**. Describe the filtrate. What do you think the filtrate contains?

10. Use the graduated cylinder to measure 25 ml of the filtrate and pour it into the evaporating dish.

11. Place the tripod above the gas burner or alcohol heater, as shown in Figure 5. Place the evaporating dish on the tripod.

Figure 5

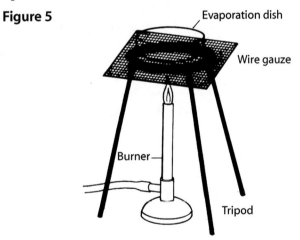

- Evaporation dish
- Wire gauze
- Burner
- Tripod

12. (⬛) **This step must be done under the supervision of your teacher or another adult.** Put on your goggles. Light the burner and adjust the flame. Slowly heat the filtrate so that the water boils off. Do not heat too fast, or some of the liquid could splatter. When all of the water has boiled off, you will be left with white crystals. This process is called **evaporation**. Turn off the heater and let the evaporating dish cool. Describe the crystals that remain in the evaporating dish.

(continued)

How Can We Separate the Parts of a Mixture? *(continued)*

DATA COLLECTION AND ANALYSIS

DATA TABLE 1

Observed Properties of Salt	Observed Properties of Soil

DATA TABLE 2

Observed Properties of Mixture	Observed Properties of Mixture + Water

1. Is the mixture a homogeneous mixture or a heterogeneous mixture? Justify your answer.

   _____

   _____

   _____

2. Describe the appearance of the decanted liquid. _____

   _____

3. Describe the filtrate. _____

   _____

4. Describe the crystals formed by evaporation of the filtrate. _____

   _____

5. What do you think the crystals are? What reasons do you have for this statement? _____

   _____

   _____

   _____

(continued)

How Can We Separate the Parts of a Mixture? *(continued)*

❓ CONCLUDING QUESTIONS

1. Describe the properties of the soil used to separate it from the mixture. _____

    _____

    _____

    _____

    _____

2. Describe the properties of the salt used to separate it from the mixture. _____

    _____

    _____

    _____

    _____

3. The techniques used in separating the mixture were decantation, filtration, and evaporation. For each of the following mixtures, explain which of these techniques you could use to separate the mixture into its components.

    (a) Coffee _____

    (b) Sugar water _____

    (c) Starch in water _____

    (d) Oil and vinegar _____

🧪 Follow-up Activities 🧪

1. Research the techniques used in water purification.
2. Research the use of distillation to separate mixtures.
3. Research the technique of fractional crystallization.

Why Are Pickles Sour?

✔ INSTRUCTIONAL OBJECTIVES

Students will be able to

- identify the properties of acids.
- observe and record data.

🌐 NATIONAL SCIENCE STANDARDS ADDRESSED

Students demonstrate an understanding of

- the properties of an acid.

Students demonstrate scientific inquiry and problem-solving skills by

- using physical science concepts to explain observations.
- working in teams to collect and share information.

Students demonstrate effective scientific communication by

- explaining scientific concepts to other students.

Students demonstrate competence with the tools and technologies of science by

- using laboratory equipment.
- using acid-base indicators.

✂ MATERIALS

- Medicine dropper
- 250-ml beaker
- 10-ml graduated cylinder
- Stirring rod
- Three test tubes
- Test-tube rack
- 🖐 Sulfur candle
- 🖐 Matches
- Vinegar
- 🖐 1M hydrochloric acid (HCl, aq)
- 🖐 0.2M sodium hydroxide (NaOH, aq)
- Zinc strip, 2 cm by 7 cm
- Piece of aluminum foil, 2 cm by 7 cm
- Red and blue litmus paper
- Phenolphthalein in dropper bottle
- Hydrion paper
- Goggles

🖐 = **Safety icon**

HELPFUL HINTS AND DISCUSSION

Time frame: Two class periods
Structure: Cooperative learning groups of two to four students
Location: In class

In this activity, students will investigate the properties of acids. 🖐 **Because they are working with acids, students must wear goggles, and you, or another adult, must supervise this activity.** It is safe for the students to taste the acid provided it is very dilute. If you do not have a sulfur candle, the students can be instructed to burn a small amount of flowers of sulfur on a piece of tin.

🖐 **The burning of sulfur must be done in a well-ventilated room and under a teacher's supervision.**

ADAPTATIONS FOR HIGH AND LOW ACHIEVERS

High Achievers: These students should work with lower achievers and assist them. Encourage these students to perform the additional activities, particularly follow-up activities 1 and 2.

Low Achievers: Challenged students may need help and guidance in performing these activities. It is important to emphasize appropriate safety precautions.

SCORING RUBRIC

Full credit should be given to students who record observations and provide correct answers in full sentences to all the questions. Extra credit may be given for completion of any follow-up activities.

INTERNET TIE-INS
    www.lhs.berkeley.edu/actacidbase.html
    http://qlink.queensu.ca/~6jem4/house.html
    www.epa.gov/acidrain/student/student2.html
    www.scsd.k12.ny.us/fowler/acid_rain_links.htm

QUIZ
    1. Why do some foods taste sour?
    2. List three properties of an acid.

Name _____   Date _____

Why Are Pickles Sour?

⚗️ BEFORE YOU BEGIN ⚗️

Have you ever eaten a pickle and wondered why it tasted sour? Can you think of other foods that taste sour? Pickles are produced by letting cucumbers sit in vinegar. It is the acetic acid in the vinegar that gives them their sour taste. In fact, all acids have a sour taste. Your tongue has chemical receptors that recognize the presence of an acid by its sour taste. In this activity, you will investigate the properties of acids.

Some examples of acids and their formulas are hydrochloric acid (HCl), nitric acid (HNO_3), sulfuric acid (H_2SO_4), and acetic acid ($HC_2H_3O_2$). Look at their formulas and find an element common to all of them. All acids contain hydrogen and release hydrogen ions (H^+) in water. Ions are charged atoms or groups of atoms. The hydrogen ion has a positive charge, which is why it is written H^+. Hydrogen ions give acids their typical properties.

In this activity, you will use several **acid-base indicators** to detect hydrogen ions. These are chemicals that change color in the presence of acids or bases. Sometimes, pieces of paper are soaked in an indicator to make a test paper. The indicators you will use are litmus paper, Hydrion paper, and phenolphthalein.

✂️ MATERIALS

- Medicine dropper
- 250-ml beaker
- 10-ml graduated cylinder
- Stirring rod
- Three test tubes
- Test-tube rack
- 🖐️ Sulfur candle
- 🖐️ Matches
- Vinegar

- 🖐️ 1M hydrochloric acid (HCl, aq)
- 🖐️ 0.2M sodium hydroxide (NaOH, aq)
- Zinc strip, 2 cm by 7 cm
- Piece of aluminum foil, 2 cm by 7 cm
- Red and blue litmus paper
- Phenolphthalein in dropper bottle
- Hydrion paper
- Goggles

🖐️ = Safety icon

PROCEDURES

Record all of your observations and answers to questions in the Data Collection and Analysis section.

🖐️ **You must wear goggles while performing this activity.**

1. Fill the beaker until it is approximately half full with water. Using the medicine dropper, add 20 drops of hydrochloric acid to the water and stir with the stirring rod.

2. Use the stirring rod to transfer a drop of this very dilute acid to a finger tip. Then, taste it. Describe the taste of the acid.

3. Use the stirring rod to transfer a few drops of the dilute hydrochloric acid to a piece of blue litmus paper. Describe what happens.

(continued)

Why Are Pickles Sour? *(continued)*

4. Rinse the stirring rod. Then, use it to transfer a few drops of vinegar to a piece of blue litmus paper. Describe what happens.

5. Rinse the stirring rod and use it to transfer a few drops of the dilute hydrochloric acid to a piece of Hydrion paper. Describe what happens.

   ✋ **Note: Steps 6, 7, 8, 9, 10, and 11 must be performed under the supervision of your teacher or another adult.**

6. Place the zinc strip in a test tube. Use the graduated cylinder to measure 5 ml of hydrochloric acid and add it to the test tube. Describe what happens.

7. Place the strip of aluminum foil in a test tube. Use the graduated cylinder to measure 5 ml of hydrochloric acid and add it to the test tube. Describe what happens.

8. Wash and dry the graduated cylinder before adding 5 ml of sodium hydroxide (NaOH) to a test tube. Use a clean medicine dropper to add two drops of phenolphthalein to the test tube containing NaOH. Describe the color of this solution.

9. Wash the graduated cylinder before using it to measure 3 ml of hydrochloric acid. Then, use a clean medicine dropper to add the hydrochloric acid, one drop at a time, to the test tube containing NaOH and phenolphthalein. Describe what happens.

10. Use a clean medicine dropper to add a few drops of NaOH to the test tube prepared in step 9 and observe the change in color. Describe what happens.

11. Light the sulfur candle. Carefully hold a piece of moistened blue litmus paper over the flame. Describe what happens.

📏 DATA COLLECTION AND ANALYSIS

1. Describe the taste of the dilute hydrochloric acid. _____

   _____

2. Describe what happens to the litmus paper when hydrochloric acid is added to it.

   _____

3. Describe what happens to the litmus paper when vinegar is added to it. _____

   _____

4. What was the color of the Hydrion paper when hydrochloric acid was added to it?

   _____

5. Describe what happens to the zinc when hydrochloric acid is added to it. _____

   _____

6. Describe what happens to the aluminum when hydrochloric acid is added to it. _____

   _____

7. What was the color of phenolphthalein in sodium hydroxide? _____

   _____

(continued)

Name _____   Date _____

Why Are Pickles Sour? *(continued)*

8. What happened to the color of the phenolphthalein when the acid was added to it? _____

   _____

9. Was the color change permanent? Explain. _____

   _____

10. What happened to the blue litmus paper when exposed to the fumes of the sulfur candle?

    _____

❓ CONCLUDING QUESTIONS

1. Why do you think that pickles and lemon juice taste sour? _____

   _____

2. Make a general statement about the taste of acids. _____

   _____

3. Litmus paper and Hydrion paper contain acid-base indicators. How do acids affect these acid-base indicators? _____

   _____

4. What happens when metals, such as zinc and aluminum, react with acids? _____

   _____

5. When the sulfur candle burns, it produces sulfur dioxide. What happens when sulfur dioxide combines with water? _____

   _____

6. Write a paragraph that describes the properties of acids. _____

   _____

   _____

   _____

   _____

⚗️ Follow-up Activities ⚗️

1. Investigate and report to the class on acid rain.
2. Test solutions of nonmetallic oxides to verify that they are acids. Report your results to your teacher.
3. Investigate and report on the history of pickling food to your classmates.

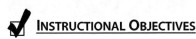

Why Are Bases Slippery?

✔ INSTRUCTIONAL OBJECTIVES

Students will be able to
- identify the properties of bases.
- observe and record data.

🌐 NATIONAL SCIENCE STANDARDS ADDRESSED

Students demonstrate an understanding of
- the properties of bases.

Students demonstrate scientific inquiry and problem-solving skills by
- using physical science concepts to explain observations.
- working in teams to collect and share information.

Students demonstrate effective scientific communication by
- explaining scientific concepts to other students.

Students demonstrate competence with the tools and technologies of science by
- using laboratory equipment.
- using acid-base indicators.

✂ MATERIALS

- Two 250-ml beakers
- 10-ml graduated cylinder
- Medicine dropper
- Stirring rod
- Three test tubes
- Test-tube rack
- Test-tube holder
- 🖐 Bunsen burner
- Fat or vegetable oil
- 🖐 Matches
- 🖐 1M hydrochloric acid (HCl, aq)
- 🖐 1M sodium hydroxide (NaOH, aq)
- Ammonia water
- Red and blue litmus paper
- Phenolphthalein in dropper bottle
- Hydrion paper
- Goggles

🖐 = Safety icon

HELPFUL HINTS AND DISCUSSION

Time frame: Two class periods
Structure: Cooperative learning groups of two to four students
Location: In class

In this activity, students will investigate the properties of bases. 🖐 **Because they are working with bases, students must wear goggles and you, or another adult, must supervise this activity.** It is safe for the students to feel small amounts of dilute bases. However, they must wash their hands immediately after feeling the base solution.

ADAPTATIONS FOR HIGH AND LOW ACHIEVERS

High Achievers: These students should work with lower achievers and assist them. Encourage these students to perform the additional activities, particularly follow-up activities 2 and 3.

Low Achievers: Challenged students may need help and guidance in performing these activities. It is important to emphasize appropriate safety precautions.

SCORING RUBRIC

Full credit should be given to students who record observations and provide correct answers in full sentences to all the questions. Extra credit may be given for completion of any follow-up activities.

💻 INTERNET TIE-INS

www.lhs.berkeley.edu/actacidbase.html
http://qlink.queensu.ca/~6jem4/house.html

❓ QUIZ

1. Why do ammonia cleaners feel slippery?
2. State three properties of a base.

Name _____   Date _____

Why Are Bases Slippery?

⚗ BEFORE YOU BEGIN ⚗

Title notwithstanding, this activity is not about sliding into second base. The bases you will investigate in this activity are a group of compounds with similar chemical and physical properties. Did you ever use ammonia to clean something? How did it feel? Ammonia cleaners contain ammonium hydroxide (NH_4OH), which is a base. In this activity, you will learn about this base and others.

Some examples of bases and their formulas are sodium hydroxide (NaOH), calcium hydroxide ($Ca(OH)_2$), and magnesium hydroxide ($Mg(OH)_2$). Look at their names; what do they all have in common? The names of these bases all end in the word *hydroxide*. A base is a compound that dissolves in water and releases hydroxide ions. The hydroxide ion is OH^-. **Ions** are charged atoms or groups of atoms. The hydroxide ion is negatively charged, which is why it is written OH^-. The hydroxide ion gives bases their characteristics.

There are several ways of testing for the presence of a base. You can use **acid-base indicators**, which are chemicals that change color in the presence of acids or bases. Sometimes, pieces of paper are soaked in an indicator to make a test paper. The indicators you will use are litmus paper, Hydrion paper, and phenolphthalein.

✂ MATERIALS

- Two 250-ml beakers
- 10-ml graduated cylinder
- Medicine dropper
- Stirring rod
- Three test tubes
- Test-tube rack
- Test-tube holder
- ⓢ Bunsen burner
- Fat or vegetable oil
- ⓢ Matches

- ⓢ 1M hydrochloric acid (HCl, aq)
- ⓢ 1M sodium hydroxide (NaOH, aq)
- Ammonia water
- Red and blue litmus paper
- Phenolphthalein in dropper bottle
- Hydrion paper
- Goggles

ⓢ = Safety icon

◈ PROCEDURES

Record all of your observations and answers to questions in the Data Collection and Analysis section.

ⓢ **You must wear goggles while performing this activity.**

1. Fill the beaker until it is approximately half full with water. Using the medicine dropper to add 20 drops of sodium hydroxide to the water. Stir thoroughly with the stirring rod.

2. Use the stirring rod to transfer a drop of this very dilute base to one finger. Then rub your fingers together. Wash your hand with running water. Describe the feel of the base.

3. Repeat step 2 using the ammonia.

(continued)

70   *Walch Hands-on Science Series: The ABCs of Chemistry*

Why Are Bases Slippery? *(continued)*

4. Using the stirring rod, transfer a few drops of the sodium hydroxide to a piece of blue litmus paper. Describe what happens.

5. Rinse the stirring rod. Then, use it to transfer a few drops of ammonia to a piece of blue litmus paper. Describe what happens. Rinse the stirring rod after you have used it.

6. Add a small amount of fat or oil to a test tube. Use the graduated cylinder to add 10 ml of sodium hydroxide to the test tube containing the fat or oil.

   🔥 **Note: Steps 7, 8, 9, and 10 must be performed under the supervision of your teacher or another adult.**

7. Light the Bunsen burner. Use the test-tube holder to hold the test tube. Heat the mixture of fat and base for five minutes. Stir the contents of the test tube while heating it. Describe the appearance of the contents of the test tube.

8. Carefully, turn off the burner and place the test tube into the test-tube rack to allow it to cool. After it has cooled, add 10 to 20 ml of water to the test tube and shake it thoroughly. Then, describe the appearance of the contents of the test tube.

9. Wash the graduated cylinder before adding 5 ml of hydrochloric acid (HCl) to a test tube. Wash the medicine dropper before using it to add 2 drops of phenolphthalein to the test tube containing HCl. Describe the color of this solution.

10. Wash the graduated cylinder. Then, measure 6 ml of sodium hydroxide. Use a clean medicine dropper to add the sodium hydroxide, one drop at a time, to the test tube of HCl. Swirl the test tube after each drop and continue adding NaOH until you observe a permanent color change. Describe what happens.

📏 **DATA COLLECTION AND ANALYSIS**

1. Describe the feel of the sodium hydroxide and ammonium hydroxide. _____

_____

2. Describe what happens to the litmus paper when sodium hydroxide is added to it. _____

_____

3. Describe what happens to the litmus paper when ammonia is added to it. _____

_____

4. What color was the color of the Hydrion paper after sodium hydroxide was added to it?

_____

5. Describe what happens when fat or oil and sodium hydroxide are heated together. _____

_____

6. Describe what happens to the test-tube contents when mixed with water and shaken. _____

_____

7. What was the color of phenolphthalein in the acid? _____

_____

(continued) 🔥

Why Are Bases Slippery? *(continued)*

8. Describe what happens when sodium hydroxide is added to the test tube containing acid and phenolphthalein. _____

_____

❓ CONCLUDING QUESTIONS

1. Why do you think that bases feel slippery? _____

_____

2. Litmus paper and Hydrion paper contain acid-base indicators. How do acids affect these acid-base indicators? _____

_____

3. What happens when bases are heated with fats or oils? _____

_____

4. Acids contain H^+ ions and bases contain OH^- ions. What do you think happens when an acid and a base are combined? _____

_____

5. Drain cleaners, such as Drano™ and Liquid Plumber™, contain bases. Why do you think bases are used to clean drains? _____

_____

6. Write a paragraph that describes the properties of bases. _____

_____

_____

_____

_____

7. Prepare a chart that compares acids and bases.

🧪 Follow-up Activities 🧪

1. Investigate and report on the manufacture of soap.
2. Plan an experiment to test household products that you think may contain bases.
3. Investigate and report to your classmates on the relationship between metal oxides and bases.

Are There Acids in the Attic and Bases in the Basement?

✔ INSTRUCTIONAL OBJECTIVES

Students will be able to

- identify various common household substances as being acids or bases.
- record data.
- use pH to identify substances as acids or bases.

🌐 NATIONAL SCIENCE STANDARDS ADDRESSED

Students demonstrate an understanding of

- the uses of acids and bases.
- pH.
- relationship between science and technology.

Students demonstrate scientific inquiry and problem-solving skills by

- using physical science concepts to explain observations.
- working individually and in teams to collect and share information.

Students demonstrate effective scientific communication by

- explaining scientific concepts to other students.

Students demonstrate competence with the tools and technologies of science by

- using laboratory equipment and acid-base indicator paper.

✂ MATERIALS

- 10 small beakers
- 10 stirring rods
- 10-ml graduated cylinder
- Scoopula
- Labels
- Marking pencil
- Window cleaner
- Club soda
- Milk
- Lemon juice
- Mouthwash
- Shampoo
- Laundry detergent
- Dishwashing liquid
- Household cleaner (e.g., Mr. Clean™, Top Job™)
- Aspirin
- Distilled water
- Red and blue litmus paper
- Hydrion paper (pH range 1–11)
- Goggles

HELPFUL HINTS AND DISCUSSION

Time frame: One class period
Structure: Individuals or cooperative learning groups of two to four students
Location: In class or at home

In this activity, students will determine the pH of some common household substances. They will be asked to generalize about their findings and should be able to determine that acidic substances are likely to be foods, while basic substances are likely to be cleaning agents. You should elicit the reasons for this from the students. If you do this activity in school under the supervision of an adult, you may wish to add drain cleaner and tile cleaner. Tile cleaners have low pHs, while drain cleaners have high pHs. **🖐Appropriate safety precautions must be followed when testing these substances.** Another variation of this activity is to use a universal indicator in place of Hydrion paper. You would need to use a color chart or samples of pH-standard solutions at a full range of pHs. Stress the importance of keeping the stirring rods for each substance separate and uncontaminated. **🖐 Because they are working with acids and bases, students must wear goggles when doing this activity.**

ADAPTATIONS FOR HIGH AND LOW ACHIEVERS

High Achievers: These students should work with lower achievers and assist them. Encourage these students to perform the additional activities, particularly follow-up activities 2 and 3.

Low Achievers: Challenged students may need help and guidance in performing these activities. It is important to emphasize appropriate safety precautions.

INTERNET TIE-INS   www.lhs.berkeley.edu/actacidbase.html
http://qlink.queensu.ca/~6jem4/house.html
http://chemistry.rutgers.edu/gemcjhem/pH.html

QUIZ   1. Why are many household cleaners basic?
2. What does a pH of 3 indicate about a substance?

Are There Acids in the Attic and Bases in the Basement?

⚗️ **BEFORE YOU BEGIN** ⚗️

By asking if there are acids in the attic and bases in the basement, we are really asking if there are acids and bases in your home. In this activity, you will use an **acid-base indicator** to determine if some common household substances are acids or bases. First, you will use red and blue litmus paper to determine which substances are acids and which are bases. Remember, red litmus turns blue in the presence of a base and blue litmus turns red in the presence of an acid. You will also find the **pH** of these substances in solution by using Hydrion paper and a color chart.

You have probably come across the term *pH,* since many products refer to it—for example, shampoo and antacid commercials. The pH of a solution is a measure of the **acidity** or **basicity** of a solution. A pH below 7 indicates that a solution is an acid. A pH above 7 indicates that a solution is a base. A pH of 7.0 is neutral, neither acid nor base. The lower the pH of a solution, the more acidic it is; the higher the pH, the more basic it is.

You will also be asked to **generalize** about household products that are likely to be acidic or basic.

✂️ MATERIALS

- 10 small beakers
- 10 stirring rods
- Labels
- Marking pencil
- 10-ml graduated cylinder
- Scoopula
- Window cleaner
- Club soda
- Milk
- Lemon juice

- Mouthwash
- Shampoo
- Laundry detergent
- Dishwashing liquid
- Household cleaner (e.g., Mr. Clean™, Top Job™)
- Aspirin
- Distilled water
- Red and blue litmus paper
- Hydrion paper (pH range 1–11)
- Goggles

🔷 PROCEDURES

🚫 **You must wear goggles while performing this activity.**

1. Place labels on each of the beakers and label them as follows: window cleaner, club soda, milk, lemon juice, mouthwash, shampoo, laundry detergent, dishwashing liquid, cleanser, aspirin.

2. Be sure to rinse the graduated cylinder after each use. Add 10 ml of window cleaner to the beaker labeled window cleaner. Continue to add 10 ml of each of the other liquids to the appropriately labeled beakers.

3. If you are using powdered laundry detergent, use the scoopula to transfer a small amount of laundry detergent to the beaker labeled laundry detergent. Add enough distilled water to half fill the beaker and stir it.

(continued) 🔥

Are There Acids in the Attic and Bases in the Basement? *(continued)*

4. Place an aspirin tablet in the beaker labeled aspirin and half fill that beaker with water. Stir to dissolve the aspirin.

5. Add approximately 10 ml of distilled water to the beakers labeled shampoo and household cleaner. Do the same to the beaker of laundry detergent if you are using liquid laundry detergent.

6. Place a piece of red litmus paper and a piece of blue litmus paper in front of each beaker. Using a separate, clean stirring rod for each substance, transfer a few drops of the contents of each beaker to the red and blue litmus paper. Determine which substances are acids and which substances are bases. Record your observations in the Data Table found in the Data Collection and Analysis section.

7. Now place a piece of Hydrion paper in front of each beaker. Again, using the same stirring rods for each substance that you used in step 6, transfer a few drops of each solution to a separate piece of Hydrion paper. Record your observations in the Data Table found in the Data Collection and Analysis section.

DATA COLLECTION AND ANALYSIS

Substance	Acid or Base (Litmus Paper)	Color of Hydrion Paper	pH
Window cleaner			
Club soda			
Milk			
Lemon juice			
Mouthwash			
Shampoo			
Laundry detergent			
Dishwashing liquid			
Cleanser			
Aspirin			

(continued)

Are There Acids in the Attic and Bases in the Basement? *(continued)*

CONCLUDING QUESTIONS

1. List all of your acidic solutions in order, starting with the least acidic and progressing to the most acidic. _____
_____

2. List all of your basic solutions in order, starting with the least basic and progressing to the most basic. _____
_____

3. What types of household substances tend to be basic? Justify your answer. _____
_____
_____
_____

4. What types of household substances tend to be acidic? Justify your answer. _____
_____
_____
_____

5. Some people, after shampooing their hair, rinse it with lemon juice or rainwater. Why might they do this? _____
_____

6. Some shampoo labels state that the shampoo is "acid balanced." What do you think "acid balanced" means? _____
_____

7. The label on oven cleaners tells you to wear rubber gloves and to work in a ventilated area. What reason can you give for this warning? _____
_____

8. On a separate piece of paper, write a brief article for the school newspaper explaining the properties of acids and bases that make them useful household products.

⚗ Follow-up Activities ⚗

1. Design an experiment to test the pH of various shampoos and hair conditioners.
2. Test other foods, such as cola soda, ginger ale, sour cream, tomato juice, fruit drinks, and yogurt to determine their acidity or basicity. Add these to the table in the Data Collection and Analysis section.
3. Research and write a report about pH meters. Read your report to the class.

Are There Indicators in Nature?

✔ INSTRUCTIONAL OBJECTIVES

Students will be able to

- identify plant extracts as indicators.
- observe and record data.
- use natural indicators to identify acidic and basic substances.

🌐 NATIONAL SCIENCE STANDARDS ADDRESSED

Students demonstrate an understanding of

- indicators.

Students demonstrate scientific inquiry and problem-solving skills by

- using physical science concepts to explain observations.
- working individually and in teams to collect and share information.

Students demonstrate effective scientific communication by

- explaining scientific concepts to other students.

Students demonstrate competence with the tools and technologies of science by

- using laboratory equipment to extract pigments.

✂ MATERIALS

- Red cabbage
- Distilled water
- 🖐 Knife
- 🖐 Blender or grater
- Stirring rod
- Funnel
- Filter paper
- 500-ml beaker (two beakers if not using a blender)
- Ring stand and ring
- Six test tubes
- Test-tube rack
- Medicine dropper
- Club soda
- Ammonia
- Lemon juice
- White vinegar
- Baking soda
- Liquid detergent
- Scoopula
- 10-ml graduated cylinder
- 100-ml graduated cylinder
- Marking pencil
- Goggles

🖐 = Safety icon

HELPFUL HINTS AND DISCUSSION

Time frame: Two class periods
Structure: Individuals or cooperative learning groups of two to four students
Location: In class or at home

   In this activity, students will make a red cabbage extract and use it to determine if some common household substances are acidic or basic. This activity can be done very well at home. Students can use clear plastic cups in place of test tubes and coffee filters and a measuring cup in place of graduated cylinders. If the students are using a blender, cut pieces of red cabbage should be combined with water to make the extract. If a blender is not available, the students can grate the cabbage and let it steep in hot water for about 20 minutes. While using the blender will produce a more concentrated extract, steeping grated cabbage in hot water will still produce an adequately concentrated solution. 🖐 **You, or another adult, must carefully supervise the use of the knife, blender, or grater.** It is important that the test substances are colorless or only slightly colored because colored solutions may mask the indicator color.

ADAPTATIONS FOR HIGH AND LOW ACHIEVERS

   High Achievers: These students should work with lower achievers and assist them. Encourage these students to perform the additional activities, particularly follow-up activities 2 and 3.

   Low Achievers: Challenged students may need adult supervision in performing this activity. It is important to emphasize appropriate safety precautions.

SCORING RUBRIC

Full credit should be given to students who record observations and provide correct answers in full sentences to all the questions. Extra credit may be given for completion of any follow-up activities.

 **INTERNET TIE-INS**   www.lhs.berkeley.edu/actacidbase.html
http://qlink.queensu.ca/~6jem4/house.html
http://chemistry.rutgers.edu/gemcjhem/pH.html
http://agrss.sherman.hawaii.edu/staff/hue/acid.html

QUIZ   1. Why do gardeners add acidic or basic substances to the soil of some flowering plants?
2. After boiling red cabbage, the cook washed the pot containing some cabbage juice with an ammonia cleaner. The contents of the pot turned green. How can you account for this?

Name _____   Date _____

Are There Indicators in Nature?

🧪 BEFORE YOU BEGIN 🧪

Gardeners sometimes add lime (a basic substance) or aluminum sulfate (an acidic solution) to the soil when growing some plants, like hydrangeas. The lime will cause the hydrangea to produce a pink flower, and the aluminum sulfate will cause it to produce blue flowers. Hydrangeas planted in neutral soil will produce white flowers. In this activity you will investigate the reason for this phenomenon.

We think of acid-base indicators as chemicals and assume that they are man-made. Actually, many acid-base indicators are plant products. For example, litmus, which is one of the most common indicators, is lichen extract. The *mus* part of the word *litmus* comes from a word that means "moss." Some early chemists, including Robert Boyle in 1664, recognized that plant extracts changed color in acids and bases. Early definitions of acids included reference to this property.

In this activity, you will make an extract of red cabbage and use it to test various acidic and basic substances found in the home. You will observe the color of the cabbage extract in these substances.

✂️ MATERIALS

- Red cabbage
- Distilled water
- 🖐 Knife
- 🖐 Blender or grater
- Stirring rod
- Funnel
- Filter paper
- 500-ml beaker (two beakers if not using a blender)
- Ring stand and ring
- Six test tubes
- Test-tube rack
- Medicine dropper

- Club soda
- Ammonia
- Lemon juice
- White vinegar
- Baking soda
- Liquid detergent
- Scoopula
- 10-ml graduated cylinder
- 100-ml graduated cylinder
- Marking pencil
- Goggles

🖐 = Safety icon

🔶 PROCEDURES

Record all of your observations and answers to questions in the Data Collection and Analysis section.

🖐 **You must wear goggles while performing this activity.**

PART I: RED CABBAGE EXTRACT

There are two different methods you can use to prepare a red cabbage extract, depending on whether you have a blender or food processor available. If you have a blender available, use Procedure A; if you don't, use Procedure B.

(continued)

Are There Indicators in Nature? *(continued)*

✋ **Note: Both of these procedures must be done under the supervision of the teacher or another adult.**

Procedure A

1. Use approximately one quarter of a small red cabbage. Cut the quarter of red cabbage into small pieces and add them to the blender. Add approximately 100 ml of water to the blender. Cover the blender and turn it on until all of the cabbage is finely ground. Turn off the blender.

2. Set up a funnel and filter paper in a ring stand above the beaker. If you are at home, you can use a coffee filter. Pour the contents of the blender into the filter and let the extract collect in the beaker below. Discard the solid material.

Procedure B

1. Use approximately one quarter of a small red cabbage. Use a grater to grate the red cabbage into small pieces. Place the cabbage pieces into a beaker or measuring cup. Using an oven mitt or potholder, pour approximately 100 ml of hot water (**not boiling**) over the cut cabbage and let it steep for approximately 20 minutes.

2. Set up a funnel and filter paper in a ring stand above the second beaker. If you are at home, you can use a coffee filter. Pour the contents of the beaker into the filter and let the extract collect in the second beaker as shown at right.

PART II: TESTING THE RED CABBAGE EXTRACT

1. Place the test tubes in the test-tube rack and label them 1 through 6.

2. Use the 10-ml graduated cylinder to add 10 ml of lemon juice to test tube 1 and then rinse the graduated cylinder.

3. As you did in step 2, add 10 ml of vinegar to test tube 2, 10 ml of club soda to test tube 3, and 10 ml of ammonia to test tube 4. Be sure to rinse the graduated cylinder between each use.

4. Use the scoopula to add a small amount of baking soda to test tube 5. Add 10 ml of water to this test tube and shake it well to mix the baking soda and water.

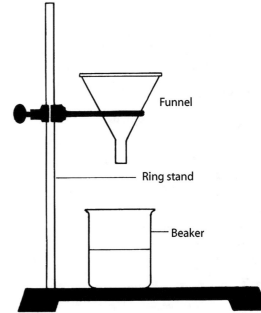

Funnel

Ring stand

Beaker

5. Add 1 ml of liquid detergent and 9 ml of water to test tube 6. Shake the test tube thoroughly.

6. Using the medicine dropper, add 10 drops of cabbage extract to each test tube. Shake each test tube thoroughly to mix its contents. Observe the color of each tube and record it in the Data Table.

(continued)

Name _____   Date _____

Are There Indicators in Nature? *(continued)*

DATA COLLECTION AND ANALYSIS

PART I

Describe the procedure that you used to produce the red cabbage extract. _____

_____

_____

PART II

DATA TABLE

Substance Tested	Color of Extract
Test Tube 1: Lemon juice	
Test Tube 2: Vinegar	
Test Tube 3: Club soda	
Test Tube 4: Ammonia	
Test Tube 5: Baking soda	
Test Tube 6: Detergent	

CONCLUDING QUESTIONS

1. Lemon juice, vinegar, and club soda are acidic. What colors do you see in these acidic solutions?

   _____

2. Ammonia, baking soda, and detergent are basic. What colors do you see in these basic solutions?

   _____

3. Why do gardeners add lime or sulfate to the soil of some plants? _____

   _____

⚗ Follow-up Activities ⚗

1. Soak a piece of filter paper in the cabbage extract and permit it to dry. Using a clean brush for each substance, paint a design on the filter paper with lemon juice, ammonia, dilute sodium hydroxide, and other acidic and basic solutions. Write a brief article describing how you did this and explaining how it works.

2. Working with pH standard solutions at pH 1 through 13, combine 10 ml of each standard with 5 drops of cabbage extract. Prepare a chart showing the color of cabbage extract at each pH.

3. Prepare extracts of other vegetable products chosen from the list below and test them to determine the color of each extract in different acids and bases. You can produce extracts of blueberries, cherries, red onion, beets, rhubarb, day lilies, roses, black tea, and violets.

Why Isn't Our Blood Like an Alien's?

✔ INSTRUCTIONAL OBJECTIVES

Students will be able to

- explain the action of buffers.
- observe and record data.
- identify practical uses of buffers.

🌐 NATIONAL SCIENCE STANDARDS ADDRESSED

Students demonstrate an understanding of

- buffers.
- pH.

Students demonstrate scientific inquiry and problem-solving skills by

- using physical science concepts to explain observations.
- working individually and in teams to collect and share information.

Students demonstrate effective scientific communication by

- explaining scientific concepts to other students.

✂ MATERIALS

- Four 250-ml small beakers
- Four stirring rods
- 100-ml graduated cylinder
- Marking pencil
- Two medicine droppers
- Distilled water
- Phenolphthalein in dropper bottle
- Methyl orange in dropper bottle
- Buffer solution with a pH of 7
- 🖐 0.1 M sodium hydroxide (NaOH)
- 🖐 0.1 M hydrochloric acid (HCI)
- Goggles

🖐 = Safety icon

HELPFUL HINTS AND DISCUSSION

Time frame: One or two class periods
Structure: Individuals or cooperative learning groups of two to four students
Location: In class

In this activity, students will observe the effect of buffers. They will observe and compare the effect of adding first a base, and then an acid, to water and to a buffered solution. 🖐 **Because they are working with acidic and basic materials, students should wear goggles for this activity.**

If you do not have commercially prepared buffers, you can prepare the pH 7 buffer by dissolving 7.0 g of potassium acid phosphate ($K_2H_2PO_4$) in 295 ml of 0.1M potassium hydroxide (KOH) and diluting to 1 liter of solution. You can substitute other indicators for methyl orange (pK_a =3.8) and phenolphthalein (pK_a =9.2), provided the pK_a of the substituted indicator is approximately the same (±1).

ADAPTATIONS FOR HIGH AND LOW ACHIEVERS

High Achievers: You may give these students a fuller explanation of how buffers work. For example, you might explain that buffers contain substances that react with hydrogen ions or hydroxide ions, thus taking up any additional H^+ or OH^- ions that are added. Depending on the definitions of acids and bases you have taught, you may wish to discuss conjugate acid-base pairs or a buffer as the combination of a weak acid and its salt. Encourage these students to perform the additional activities, particularly follow-up activities 2 and 3.

Low Achievers: Challenged students may need help and guidance in performing these activities. It is important to emphasize appropriate safety precautions. Limit your discussion of buffers to a descriptive definition—for example, a buffer is a substance that resists change in pH when an acid or base is added to it.

SCORING RUBRIC

Full credit should be given to students who record observations and provide correct answers in full sentences to all the questions. Extra credit may be given for completion of any follow-up activities.

 **INTERNET TIE-INS**

www.lhs.berkeley.edu/actacidbase.html

http://qlink.queensu.ca/~6jem4/house.html

http://chemistry.rutgers.edu/gemcjhem/pH.html

www.epa.gov/acidrain/student/exp8.html

 QUIZ

1. What does it mean to say that a medication is buffered?
2. Why do some people use buffered aspirin?

Name _____ Date _____

Why Isn't Our Blood Like an Alien's?

⚗ BEFORE YOU BEGIN ⚗

In the movie *Alien*, the creature's blood was very acidic. Every time the creature was wounded, the acidic blood from the creature ate through the floor of the spaceship. Can you imagine our blood getting that acidic, or, for that matter, the blood one of the earthbound monsters we have seen in recent movies, such as Godzilla? If we ate a lot of sour pickles or lemon juice, would our blood become more acidic? Our blood, and that of other animals on earth, maintains a constant **pH**. The pH of a solution is a measure of the acidity or basicity of a solution. A solution with a pH below 7 is an acid, and a substance with a pH above 7 is a base.

The pH of blood is maintained in a narrow range which varies between 7.3 and 7.5. This is because blood contains substances called **buffers**. Buffers are chemicals that keep pH fairly constant even when acids or bases are added to them. You may have come across buffers in medicines. For example, many people use buffered aspirin so that the acid in aspirin does not make their stomachs too acidic.

In this activity, you will investigate how buffers work. You will do this by comparing the effect of adding a base both to water and to a buffered solution. Then, you will make the same comparison using an acid. You will use acid-base indicators to indicate changes in pH.

✂ MATERIALS

- Four 250-ml small beakers
- Four stirring rods
- 100-ml graduated cylinder
- Marking pencil
- Two medicine droppers
- Distilled water
- Phenolphthalein in dropper bottle

- Methyl orange in dropper bottle
- Buffer solution with a pH of 7
- 🖐 0.1 M sodium hydroxide (NaOH)
- 🖐 0.1 M hydrochloric acid (HCl)
- Goggles

🖐 = Safety icon

🧪 PROCEDURES

🖐 **You must wear goggles while performing this activity.**

1. Use the marking pencil to label the beakers Water A, Buffer A, Water B, and Buffer B—only one name for each beaker.

2. Use the graduated cylinder to add 100 ml of distilled water to each "Water" beaker.

3. Use the graduated cylinder to add 100 ml of buffered solution to each "Buffer" beaker.

4. Add five drops of phenolphthalein to the beaker labeled Water B and five drops of phenolphthalein to the beaker labeled Buffer B.

(continued) 🔥

Why Isn't Our Blood Like an Alien's? *(continued)*

5. Use the medicine dropper to add sodium hydroxide drop by drop to the beaker labeled Water B, stirring after each drop. Continue adding the sodium hydroxide to the beaker, counting each drop, until a permanent pink color forms. Record the number of drops it took to get a permanent pink color in the Data Table found in the Data Collection and Analysis section.

6. Use the medicine dropper to add drops of sodium hydroxide to the beaker labeled Buffer B, stirring after each drop. Continue adding sodium hydroxide to the beaker, counting each drop, until a permanent pink color forms. Record the number of drops required to get a permanent pink color in the Data Table found in the Data Collection and Analysis section.

7. Add five drops of methyl orange solution to the beaker labeled Water A and five drops of methyl orange solution to the beaker labeled Buffer A.

8. Use the second medicine dropper to add hydrochloric acid one drop at a time to the beaker labeled Water A, stirring after each drop. Continue adding hydrochloric acid to the beaker, counting each drop, until a permanent red color forms. Record the number of drops it took to get a permanent red color in the Data Table found in the Data Collection and Analysis section.

9. Use the medicine dropper to add drops of hydrochloric acid to the beaker labeled Buffer A, stirring after each drop. Continue adding hydrochloric acid to the beaker, counting each drop, until a permanent red color forms. Record the number of drops required to get a permanent red color in the Data Table found in the Data Collection and Analysis section.

DATA COLLECTION AND ANALYSIS

DATA TABLE

Solutions	Number of Drops to Obtain Color Change
Water B (water and base)	
Buffer B (buffer solution and base)	
Water A (water and acid)	
Buffer A (buffer solution and acid)	

1. How many more drops of base did you need to add to the buffer solution than to the water to change the color of the indicator?

   _____

2. How many more drops of acid did you need to add to the buffered solution than to the water to change the color of the indicator?

   _____

(continued)